INTERACTIVE VIDEO: IMPLICATIONS FOR EDUCATION AND TRAINING

Working Paper 22

INTERACTIVE VIDEO
implications for education and training

John Duke
Management, Education and Training Consultant

CET Council for Educational Technology

Published and distributed by the
Council for Educational Technology,
3 Devonshire Street, London W1N 2BA

© Council for Educational Technology 1983

First published 1983

ISBN 0 86184-105-0
ISSN 0307-9511

Other than as permitted under the Copyright Act 1956 no
part of this publication may be photocopied, recorded or
otherwise reproduced, stored in a retrieval system or
transmitted in any form by any electronic or mechanical
means without the prior permission of the copyright owner.

 British Library Cataloguing in Publication Data

Duke, John
    Interactive video. — (Working paper/Council for
Educational Technology, ISSN 0307-9511 ; 22)
    1. Video tapes in education
    I. Title   II. Series
    371.33'5   LB1044.75

ISBN 0-86184-105-0

Printed in Great Britain by
Direct Design (Bournemouth) Ltd, Printers
Sturminster Newton
Dorset DT10 1AZ

CONTENTS

	Foreword	9
1.	Background	11
2.	Scope of the study	14
3.	Relationship of interactive video to other information communication systems	17
	Educational television	17
	Computer-based learning	19
	Information/communication systems	20
4.	The unique characteristics of interactive video	22
5.	Interactive video using a random-access videotape recorder	25
	CAVIS (Computer Audio-Visual Instruction System)	28
	FELIX (Felix Learning Systems Ltd)	29
	RESPONDER (Sony UK Ltd)	30
	Microcomputer/VCR interfaces	31
6.	Optical recording media	33
	Videodisc technology	34
	Optical videodiscs	35
	LaserVision	37
7.	The LaserVision system	40
	Classification of LaserVision players	44
	Technical features of available hardware	46
8.	Description of typical systems	49
9.	Videotex and videodisc technologies	58

10.	**Systems software**	62
	Control programs	62
	Command codes	63
	Authoring software	65
	Student-oriented software	68
11.	**Databases for interactive video**	70
	Video images	70
	Still images	71
	Textual information	72
	Computer data files	73
	Audio databases	73
	Integration	74
12.	**Courseware considerations**	75
	Presentational features	76
	Instructional strategies	77
	Programme design	77
	Learning styles	79
13.	**Videodisc production**	81
	Film	82
	Slide	82
	Videotape	83
	Electronic image handling	83
	Authoring technique	84
	Editing and evaluation	85
	Mastering	86
	Production costs	87
14.	**Resource requirements**	89
	Manpower skills	89
	Equipment and facilities	90
	Computing facilities	92
	User terminals	94
15.	**Economic considerations**	95
	Scale of programme	97

CONTENTS

16.	**A model research and development programme**	101
17.	**Applications of interactive video in education and training**	104
	Schools	104
	Further and higher education	106
	Industrial and corporate training	108
	Military training	110
	Adult/recreational education	110
18.	**Potential for collaboration**	112
	Further and higher education	112
	Public corporations	114
	Industrial training	114
	Military training	116
19.	**Executive summary and recommendations**	117
	References	126
A1.	**Digital data discs**	128
	Digital read-only memory	128
	High-resolution image store	130
	Optical digital data discs	130
A2.	**LaserVision systems: provisional description of computer interface for use with the Philips prototype industrial player**	132
A3.	**LaserVision programme master tape specification (PAL 625/50] December 1982**	142

LIST OF ILLUSTRATIONS

Figure 4.1. A paradigm for presentational media, after Copeland, 1981.	23
Figure 5.1. Schematic interface between microcomputer and VCR	26
Table 6.1. Cost comparison of storage media	34
Figure 7.1. Modulation of video signal and subsequent limiting	41
Figure 7.2. Recording and playback processes — schematic	42
Figure 7.3. Layout of frame information on CAV and CLV discs	45
Table 7.1. Comparison of NTSC players	47
Table 7.2. Comparison of PAL players	48
Figure 8.1. Pulse-code modulation of VP1000 control signals	50
Table 8.1. Pioneer VP1000 control codes	51
Table 10.1. Control codes for Pioneer computer/videoplayer interface	64
Table 10.2. Command table for prototype Philips industrial player	66
Figure A1.1. NTSC colour video signal	129
Figure A3.1. Preferred format for all tapes	148

FOREWORD

This book is a result of the happy coincidence of instigator, investigator and sponsor. The subject of study, the linking of video recording systems, in particular the videodisc, with computers in education and training applications, is clearly one with which the Council for Educational Technology should be concerned, and it was the Council, as part of its information technology programme, which initiated this work. But the grant the Council receives has been reduced. It was never enough to support all the worthwhile work falling within the Council's very broad remit, and it is now even more restricted. We were fortunate to engage the interest of Rediffusion Computers Limited and the Department of Industry, and with their support we were able to put the work in hand. The third element was the availability of John Duke, lately Assistant Director at the Middlesex Polytechnic and one-time Assistant Director of the National Council for Educational Technology. The quality of his contribution is demonstrated by the following pages.

The art of educational development, for it is far from being a science, is to identify new systems and technologies which have potential in education and training at an early stage, preferably when the technology has settled into a stable and predictable development pattern but while the nature of the applications are still evolving. This we hope we have done with this report. The convergence of video and computing offers an interactive system of great flexibility. Yet it is almost a truism to say that a system is only as good as its software, and the development of education and training software for interactive video is an uncharted field. The possibilities opened up by this report are in two main areas. The first, and most apparent, is the potential of interactive video in practically all aspects of education and training. The second is that the United Kingdom could develop the capacity to create high-quality software with a world-wide market.

INTERACTIVE VIDEO

It is clear that adoption in the education and training field will move relatively slowly. The essential equipment is hardly yet available, and the even more essential software, which as Mr Duke shows is not readily produced on a cottage industry basis, is even less available. Hence, if we are to create a new sector in the knowledge industry it will have to be done in advance of home demand, and will have itself to stimulate and generate home demand.

As the report shows, any such venture would involve substantial investment over a number of years. It would involve the collaboration of a number of public and private interests. It would require a good deal of vision and steady nerves, and the prize would be substantial, both for the UK economy and for the UK education and training system.

This report is published as a first contribution to a discussion which must take place soon and swiftly if it is to lead to a satisfactory outcome. CET will be exploring with potential partners the possibility of taking the next step in this exciting new area. Meanwhile our thanks are due to the investigator for a wide-ranging and perceptive report, and to the sponsors for their support, both financial and moral.

Geoffrey Hubbard
Director, CET
April 1983

1. BACKGROUND

1.1 Considerable resources have been invested by the television and film industries in recent years to develop systems that allow the individual viewer to select his own choice of television programme to supplement the offerings of the broadcast networks. These activities have been supported on the one hand by the electronics manufacturers who have sought thereby additional markets for consumer products, and by the film and programme makers and distributors seeking further outlets for their products. A wide range of approaches has been followed, falling broadly into two categories: those based on the distribution of discrete, pre-recorded materials, and those based on alternative means of access to central or local networks capable of offering multi-channel facilities to increase programme choice. The lure of the potential market is so great that considerable investments have been attracted, and the economic and social consequences such that government intervention and regulation have become controlling factors.

1.2 Although many systems have been investigated the establishment of the magnetic-tape videocassette recorder as the dominant product in the domestic market has been an outstanding success story, with current estimates of over two-and-a-half million installed machines in the United Kingdom and a growth rate of over 50 per cent per annum. This market is largely exploited by the Japanese electronics industry, which has achieved a remarkable level of precision and reliability in product engineering and demonstrated considerable marketing flair in getting the public to accept relatively expensive systems, despite the existence of competing standards.

1.3 Within the educational and training worlds a similar uptake of videotape technology has been evidenced. The advantages of release from the limitations of scheduled programming can thereby be realized, and the local or minority interests of teaching and learning groups catered for by specially produced or adapted programmes. Even authorities such as the ILEA that originally relied on cabled distribution of educational television discontinued network services in favour of the more flexible and cost-economic videocassette technology. The outcome of renewed interest in cable distribution of educational programmes, particularly to non-institutional audiences, will depend largely on the characteristics of the chosen technology and of the economic and regulatory framework within which it is operated. Despite the predominant influence of 'professional' standards in the tertiary and industrial training sectors, the majority of schools and colleges rely on domestic standard equipment, and this factor has normally to be taken into account in any consideration of the extended use of videorecording techniques in education.

1.4 Although educational television has proved itself as a significant resource over its 20 years or more of history, a number of deficiencies can be identified that limit its effectiveness as a presentation medium. These become more acute as its use narrows from that of a mass communication device to one that serves the specific requirements of an individual learner. This is precisely the application that would appear most attractive, with its intimate screen, good quality reproduction of realistic, moving images in full colour, and ability to attract and hold the attention of the viewer. It is a fundamental assumption of this report that the realization of the full potential of television as a powerful vehicle for the support and extension of individual study methods has hitherto been hindered by technical factors, which in themselves have constrained the creative imaginations of programme authors and producers. New developments in television recording technology are beginning to make possible a more learner-oriented approach to the use of television as an instructional medium, by allowing the prospect of a marked degree of 'interactivity' between viewer and programmes, which at the limit will represent a totally individual experience for each person.

1.5 The concept of 'interaction' has evolved from the elementary ideas of programmed learning through the versatility of experiences offered by computer-based learning systems. Even these, however, despite complex and sophisticated graphics facilities, have not been able to exploit adequately and integrate the use of moving pictures to any great extent. However, the convergence of computer and video-recording technologies has now reached a stage where it is possible by computer control of a videotape player to obtain sufficient random access to recorded sequences to utilize these as active components of an individualized learning scheme. Moreover, the emerging videodisc technology promises more important freedom to present high-quality moving and still pictures in rapid random-access, or in slow motion, with great flexibility, under computer-guided sequencing. The advent of the videodisc, television's equivalent to the long-playing gramophone record, at last appears to release the television programme from the linear sequential format it inherited from the motion film; offers opportunities for constructive symbiosis between the computing and television worlds; and holds out to the educational technologist an exciting new component with which to construct interactive, individualized learning systems.

2. SCOPE OF THE STUDY

2.1 The Council for Educational Technology has a role to advise on the introduction of information technology in education and training, and through its information technology programme it is concerned to investigate not only individual technologies but also the convergence of these technologies and the opportunities that arise for the creation of new educational environments and resources. The programme conducts studies and trials, and supports specific projects to develop hardware and to produce teaching and learning materials. The results of this programme are actively disseminated to promote appropriate uses of information technology.

2.2 Interactive video exemplifies the powerful possibilities resulting from the convergence of technologies. The combination of television and computer-assisted learning techniques promises a basis for an important new educational and training tool with unique characteristics. Interactive video could become a significant component of a future 'electronic learning station', capable of making available to the individual learner a rich variety of facilities to support a wide range of educational tasks.

2.3 The development of interactive video will require specialized resources and skills to plan appropriate applications and to design and produce materials. The process demands a fusion of existing writing and production skills together with computer software expertise, and also the ability to develop a style suitable for the presentation of teaching material in this new medium. The introduction of interactive video needs to take place within the overall design of a learning system. There are also technical problems to be overcome in the preparation of material concerning the merging of different media and the structuring of sequences.

SCOPE OF THE STUDY

2.4 In addition to expertise in the design and production of material, the successful introduction of interactive video will depend on the generation of experience in selection of appropriate hardware systems, in financial costing and control, and in the training of programme producers and course organizers. Without the creation of such a supporting infrastructure there is danger that any investment in this new technology will prove fruitless due to lack of relevant material or because the material fails to fulfil the claims made for it.

2.5 The United Kingdom is recognized as a world leader in television production and in computer software. Within education and training there are considerable experience and resources in both educational uses of television and in the development of computer applications to learning. There is a growing awareness that the commercial potential of these resources has so far been underexploited, and that these could provide the basis for the creation of the new 'knowledge-based' industry with substantial export potential.

2.6 The Council for Educational Technology therefore commissioned a study with the following aims:
(i) to describe the characteristics of interactive video, its unique facilities and its relationship to the other components of an 'electronic learning station'
(ii) to identify applications for interactive video in education and training
(iii) to investigate the current state of the art and to identify potential sources of expertise for software development
(iv) to identify the skills required to develop effective materials covering course design, programme creation, and the production and compilation of material
(v) to investigate the scale of costs involved in producing interactive video materials and the cost benefits to the user
(vi) to identify pilot projects for more detailed investigation, together with appropriate outline time-scales, resource costs and potential collaborating institutions and organizations
(vii) to identify possible problems and difficulties to be overcome in developing such a programme.

INTERACTIVE VIDEO

2.7 Although 'interactive video' has a close relationship to a number of information presentation techniques, as is illustrated in Section 3 of this report, this study has concentrated on an interpretation of interactive video that encompasses the use of educational/training/library materials offering the individual user a significant degree of dialogue, or interaction, that determines the sequence of presentation, and designed for videodisc or videocassette players associated with computerized control, either through an internal microprocessor or through connection to an external computer.

2.8 The study was conducted largely through conversations with senior staff in the electronics, television and publishing industries, and with academic research workers and educational technologists. Further technical data has been derived from published information in technical journals and trade periodicals. A significant input was provided by Vincent Thompson, Assistant Director of the Council for Educational Technology, following a recent visit to a number of educational research centres in the USA. The author is grateful to all those who have provided information and advice.

3. RELATIONSHIP OF INTERACTIVE VIDEO TO OTHER INFORMATION/COMMUNICATION SYSTEMS

Educational television

3.1 The widespread familiarity of television, available in almost every home in the country, is a major advantage from an educational point of view. As well as ready access, it offers ease and convenience of use. It has powerful attraction as an entertainment medium. Such characteristics assist its acceptance as an instructional medium within a formal educational context, and are of considerable relevance in allowing informal educational experiences to be obtained, often incidentally, by the general public. The better exploitation of the educational possibilities of television, however, relies on the degrees to which individual control can be extended by the learner over when, where and how the medium is used. It is significant that the home television set is now being seen as a general-purpose display device, through the growth of video games, home computers and videotex systems, all of which offer the user a degree of control over the medium.

3.2 With broadcast television the obvious limitations are that it operates to fixed programming schedules and that educational programmes have to compete for scarce air-time with other programmes. Programmes are of an ephemeral nature and are designed as holistic units of continuous narrative or argument. There is no opportunity for the viewer to pause or reflect, or to take notes while viewing, without losing touch with the thread of the argument, and since pace, level, format, structure and language have been chosen by the producer to suit the perceived average viewer, the appropriateness of any particular programme for any particular student is more a matter of chance than of design. Since the programme originators have no direct contact with their students, and the students themselves cannot question the system to clarify their own understanding, there is a tendency towards passive acceptance as the primary learner response.

INTERACTIVE VIDEO

3.3 These characteristics are also largely common to other forms of 'broadcast' distribution such as cable networks at their present state of development, with the exception that the viewer may be able to 'dial up' a particular programme in which he is interested, and that very limited possibilities may exist for viewer responses to be fed back to the originating centre in answer to multiple-choice questions or some simple polling device — an extremely primitive form of 'interaction'.

3.4 Some of these features can be alleviated through use of recorded television, which allows, *inter alia*, freedom from scheduling constraints, and the ability to stop, slow down and retrace steps during a programme, and which encourages production of materials in segments designed to permit local restructuring to suit the characteristics of the learner as seen by his immediate teacher or counsellor. Materials for different abilities and levels of learners can be more readily made available, with a degree of control exercised by the learner over the programme solely dependent of the features of the playback equipment. The development of videotape machines which offer random access to the contents of a recording, and which may be controlled by external signals, has heralded the possibility of making independent learning from televised images a reality, although the current limitations of magnetic tape technology produce serious limitations on the comprehensiveness and flexibility of this approach.

3.5 Some interactive video systems based on the programmable videocassette recorders are described in Section 5. The advent of the optical videodisc, an alternative medium to magnetic videotape as a storage and transport facility for television recordings, provides a new dimension in the technical ability to deliver and manipulate video sequences. Such is the potential of this new technology in an educational and training context that the bulk of this report will be devoted to discussing the implications of this development.

RELATIONSHIP TO OTHER SYSTEMS

Computer-based learning

3.6 The computer has become the crucial element in all but the most simplistic interactive learning systems. Computer-managed learning (CML) systems have been designed to direct and control the individual learning pathways of groups of students working from printed worksheets and other pre-prepared materials, and to deal with the resource allocation and continuous assessment issues that are characteristic of a programmed learning approach. Many examples of computer-assisted learning (CAL) and computer-aided training (CAT) systems have been described in which the computer takes on aspects of the delivery of information to the student and simultaneously provides a channel of communication through which the student responds to the material and thereby determines the subsequent steps presented. Highly sophisticated CAL systems have been developed offering a range of educational experiences beyond simple drill and practice, including the ability to model and simulate real-life situations, and to interrogate and manipulate complex databases. Despite the introduction of extensive and high-resolution computer-generated graphics facilities a major limitation on all existing computer-based learning systems is their inability readily and realistically to control and display real-world pictures in true colour and dynamic motion.

3.7 The visual display unit (VDU) has become an almost universal terminal for all types of computer system, whether the data to be presented is alphabetic or numeric or graphic, and set out in textual, tabular or pictorial form. VDUs fall into two broad categories: CRT monitors, displaying dot matrix patterns; and raster-driven displays based on conventional broadcast television practice. The former operate primarily in digital mode, the latter in analogue mode. The former can display very high-quality, high-resolution images, in both monochrome and colour, but the need continuously to refresh the picture demands high-speed processing and a large main memory. By transforming the digital output of a computer into an RF modulated waveform, computer-generated data can be displayed on domestic television sets. This makes lesser demands on processing power, but has the disadvantage that the inherent lower resolution severely limits the size and amount of text that can be satisfactorily displayed.

3.8 Many computer-based learning systems have been designed applying both these approaches to information display, but the issue of combining computer-generated images and camera-generated video images in the same display has not yet been totally resolved. Computer-based learning has so far been limited by the inability to call up real-life moving pictures and animations (and with an associated sound track) with a speed and facility that matches other computer operations. Interactive video, particularly utilizing the features of the videodisc, offers a solution to this problem, although it will not be the only application of videodisc technology to learning system development.

Information/communication systems

3.9 The educational world is only gradually beginning to appreciate the opportunities offered by electronic storage, manipulation and dissemination of data. The ability to access and interrogate comprehensive databases can provide for both teachers and learners an important new range of educational resources. Most progress has been made with bibliographic databases that offer information about library materials, scientific and technical publications, and other professionally oriented source material. Experimentation with more general information dissemination systems such as PRESTEL has also indicated a considerable educational potential. (Videotex systems are described in detail in Section 9.) Major questions for consideration relate to the extent to which widespread access to such networks may be provided, and the costs, and the ease and extent with which the individual user may interact with the systems.

3.10 Interactive video techniques can be combined with these computerized database systems to enhance their facilities and improve their availability. The optical videodisc has very impressive storage capacity and in terms of cost per bit is

RELATIONSHIP TO OTHER SYSTEMS

six times cheaper than photographic film and 400 times cheaper than magnetic disc. It has particular advantages where portability is required, where images need to be stored (although it can also handle text satisfactorily), and where its permanent nature makes accidental erasure unlikely and offers archival quality. It could become a significant component of computerized educational information systems where there is little need for constant updating. The data handling characteristics of videodisc are more fully described in Section 6.

3.11 Other information technologies are contributing to interactive video development. Standard telephone networks are capable of distributing compressed slow-scan or freeze-frame video signals which can be reconstituted on a television screen. Devices such as the Open University's CYCLOPS take advantage of this capacity to offer electronic blackboard facilities so that teachers can communicate images to distant students, and facsimile services allow transmission of documents and pictures which may be stored and independently accessed at user stations. Viewdata services offer possibilities of limited interaction through response frame facilities, but it is perhaps the development of viewdata techniques as a bridge between computer-generated text and imagery and televised pictures that holds the best short-term promise. The use of telesoftware techniques, in both broadcast (CEEFAX, ORACLE) modes and as PRESTEL transmissions to distribute computer programs for use in combination with video material, is also of significance in considering possible directions of development of interactive video. With the expected spread of local area networks, videodisc may conceivably become an important shared resource.

4. THE UNIQUE CHARACTERISTICS OF INTERACTIVE VIDEO

4.1 Copeland (1981) describes an educational technologist's approach to the relationship between communications engineering and the process of human learning. Much teaching takes place through expository presentation, although it is well documented that the effectiveness of any teaching process is enhanced if the learner is encouraged to respond and demonstrate his degree of understanding. With teacher and learner in a one-to-one tutorial this arrangement is easy to achieve, but in situations where the communications path between teacher and taught is more extended the ability to deal with learner response is less happy. In the case of educational television the ability to induce learner response is extremely limited in that the transient nature of the medium itself offers little opportunity for the viewer to reflect in answer to a question, and there is no channel through which a response can be transmitted and acknowledged.

4.2 In assessing the properties of educational media in terms of this expositional/inquisitional continuum, it is also helpful to consider a further dimension which at one extreme includes all permanent images (such as printed text) and at the other transient displays (such as film). By use of this device Copeland suggests the presentational features of all educational media can be represented diagrammatically as in Figure 4.1.

4.3 Whereas print is permanent and expository, television is largely transient and expository; and suitable placings can be identified for most current media. It is postulated that the most efficient medium is that which contributes to all four quadrants; a feature most offered by computer-based learning.

UNIQUE CHARACTERISTICS OF INTERACTIVE VIDEO

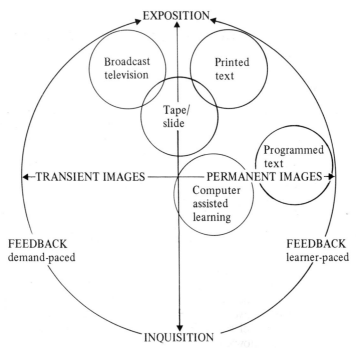

Figure 4.1 A paradigm for presentational media, after Copeland, 1981. (The Council thanks Peter Copeland of Futuremedia Ltd, 7 East Pallant, Chichester, PO19 1TR, for permission to use this material.)

4.4   A further factor, however, in aiding learning is the value of contributory elements such as prompts, aids and guidance, which, since they are often most effective when transient and accompanied by spoken and pictorial explanation so that they supplant or support the learner's own mental imagery, fall in the top-left quadrant. It is also necessary to consider the provision of feedback routes, one representing system-paced responses, and another representing learner control over the process. The features of a fully adaptive medium thus become apparent:

INTERACTIVE VIDEO

— it should have attributes that extend into all the dimensions indicated
— the way in which information is structured should allow redesign to match the study strategy of the learner
— performance data should be accumulated to assist the selective presentation of help and assistance
— it should have capacity to motivate the learner, particularly through the influence he can exercise over the teaching strategy
— must be economic and effective communication.

4.5 The medium with greatest potential to meet most, if not all, of these requirements is interactive video, which combines the rich and varietal sensory experiences offered by television with the detailed control characteristics of computer-based learning. The complementary nature of these two technologies promises that together, following recent developments in microcomputing and videorecording engineering, the closest match yet to the complete individual learning system can be achieved. Through such a system an individual learner may exercise control over content, structure and pace, and by links with local or national networks could call on a multitude of information sources to enhance its capacity or increase its adaptivity. Teachers can call upon text, sound and pictures, using computer assistance to create and produce learning materials and adapt and modify existing programs for their own purposes. This is the ideal of interactive video and its claim to being unique.

5. INTERACTIVE VIDEO USING A RANDOM-ACCESS VIDEOTAPE RECORDER

5.1 Most early experimentation in this country has involved the connexion of a programmable videocassette recorder to a microcomputer system. Several systems have been reported including those developed at the Open University (Laurillard, 1982), and at Dundee College of Technology (Bryce, 1982). Other equipment is offered commercially, aimed primarily at the industrial training market, by CAVIS-Scicon and FELIX Learning Systems. Within this category can also be included the Sony RESPONDER system, which offers more limited programming capability.

5.2 A typical system comprises a standard microcomputer, typically an APPLE II, connected to a U-Matic or industrial VHS player/recorder via a special interface, as shown schematically in Figure 5.1. Both video and computer output is directed to a standard video monitor.

5.3 The student sits at the microcomputer keyboard, watching the screen. The computer program instructs the VCR to present the first sequence of video material at the end of which control passes to the computer, which then presents commentary or question to test the student's understanding. Input from the student, which can be either in answer to a multiple-choice question or a constructed response, will trigger off an appropriate reaction, leading to a remedial or reinforcing loop, or to the next sequence of video material. All the techniques of computer-based learning may be applied, involving tutorial, practice, testing, simulation, etc, and the computer can be programmed to collect and analyse student performance and print out appropriate records.

5.4 The creation of such an interactive video teaching programme is relatively straightforward and designed to allow a teacher with little experience of computing to construct his

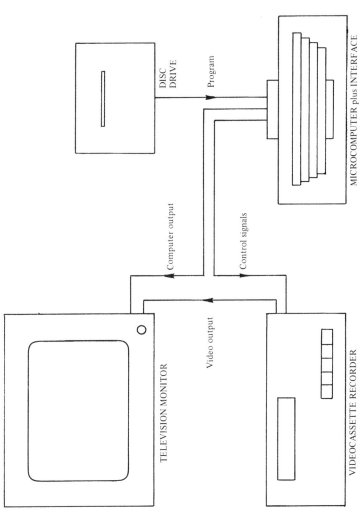

Figure 5.1. *Schematic interface between microcomputer and VCR*

USING A RANDOM-ACCESS VIDEOTAPE RECORDER

own material. The videotape is first prepared by recording a series of electronic pulses at one-second intervals on to the second audio track available on the industrial machines. This provides a framing reference by which every sequence on the tape can be uniquely specified, and which allows the computer automatically to locate any position on the tape. The computer software preferably includes an 'authoring system' that will assist the teacher to specify the content of the information to be displayed to the student between video sequences. Each 'page' of information will be accompanied by further (undisplayed) instructions to control the actions taken by the computer following the responses given by the student. The task of creating typical multiple-choice branching programs by this method is often quick and efficient.

5.5 Greater complexity can be obtained in several ways. The use of character-generation circuitry based on the viewdata standard allows simple textual comment to be superimposed on the video images. The graphic capability of the microcomputer can be utilized to display additional images, to be interspersed with the video pictures, and in certain cases superimposed on them. The computer may be programmed to accept constructed responses from the student, and subject these to numeric or textual analysis. The student may be invited to use the calculation or graphic display ability of the computer to indicate his level of understanding.

5.6 The use of videocassette tape as the medium for interactive video sequences tends to be awkward and unwieldy if more than the simplest levels of instruction are involved. The time taken for the cassette player to search for the correct frame to start a sequence can range from several to tens of seconds, particularly if the program has several branches or reentrant loops. Slow-motion facilities are limited, and stop-frame action results in rapid degradation of picture quality. VCRs are noisy in action while searching, which can cause distraction, although mechanisms are on the whole reliable. Although the VCR offers

INTERACTIVE VIDEO

an accompanying soundtrack, audio commentary to still frames requires a separate audio recorder or wastefully using the VCR with a blank vision signal.

5.7 In general, systems based on tape technology may be regarded as stop-gap solutions with a do-it-yourself emphasis. This is not to deny that valuable experience will be obtained from exploration of their abilities, and some high-quality programs will undoubtedly be produced. The temptation to take existing television footage and use it without alteration should be considered very carefully, since material designed for continuous viewing will have been derived from a different design standpoint and will either fail to represent or neglect entirely the requirements of interactivity. Failure to recognize this could discredit the medium. Further, since many different approaches and a lack of rigorous standards in the technical aspects of programme production are likely, the opportunities for exchanging courseware between systems (and even between ostensibly identical equipments) will probably be slight.

5.8 The detailed characteristics of some readily available systems are as follows.

CAVIS (Computer Audio-Visual Instruction System)
5.9 Developed by Scicon Computer Services Ltd, the system comprises a colour television monitor and student keypad. A separate editing keyboard is linked to a purpose-built microcomputer control unit and JVC industrial VHS recorder. A simple thermal printer is provided to allow hard copy of displayed data to be obtained. The system generates standard videotex format characters and graphics and this can be used to create text and simple images that can be displayed on the screen or simultaneously presented with video pictures from the VCR. Authoring software is provided on floppy disc which offers a simple command structure to assist the creation of 'pages' and control of the player.

USING A RANDOM-ACCESS VIDEOTAPE RECORDER

5.10 Commands include:

STOP	the VCR
GO	start the VCR
PAGE 21	display page 21 only
JUMP 160	jump to location 160
BOTH 25	display BOTH page 25 and VCR
AUDIO	switch audio signal on
MUTE	suppress audio signal
HOLD	hold the last event for 3 seconds
QUESTION 9	ask a question on page 9
WAIT/KP	wait for * on student keypad
RESTART	start bit earlier than GO

5.11 This straightforward command structure allows the author to create courseware with minimal computer knowledge and keyboard skills. The system is aimed at the industrial training field, on the assumption that existing training material on film or video will be modified and made interactive to suit local requirements. Up to 200 pages of videotex explanation, diagrams and questions can be stored on the floppy disc and thus quickly retrieved. The expectation, however, is that simple branching structures and multiple-choice techniques will be used, since the student keypad only offers numerical responses and a limited number of special function keys, such as Return, Skip and Index. The pattern of student responses is stored by the system for future analysis by the trainer. The system is self-contained, and costs (in December 1982) about £13,000 per station.

Felix (Felix Learning Systems Ltd)

5.12 This is another commercially available system aimed at the industrial training market, but based on the combination of a standard APPLE II microcomputer and a Sony U-matic VCR. The student workstation offers VCR and monitor, together with 48K micro, single disc-drive and full alphanumeric keyboard. Micro and VCR are connected via a special-purpose interface. The management workstation adds a second disc-drive and printer to allow operation of software to manage student/course records. The production workstation provides an editing VCR, and a more powerful

INTERACTIVE VIDEO

micro (64K), to run the authoring software and create programmes. The system was developed with the aid of a Department of Industry Microelectronics Applications Project grant, and the authoring language is based on the NPL's EDUTEXT. A student workstation markets at about £4250.

5.13 The computer controls the video presentations and also allows computer-generated material to be shown separately on the screen, but not overlaid. Since the full capacity of the micro is available to the student a range of options is open to the trainer when eliciting student responses. These can therefore include constructed responses, etc, as well as multiple-choice answers. The workstation can also be used to run standard business and management software, when it is not required for training purposes.

5.14 Although able to supply equipment to customers willing to provide their own courseware, the main concern of Felix Learning Systems Ltd is the creation of high-quality training materials. The company, part of the Viscom group, has a professional design team and comprehensive production resources, and is particularly strong in the management field.

Responder (Sony UK Ltd)

5.15 The Responder system adds simple interactive programming facilities to a random-access U-matic VCR. The student keypad offers ten numeric keys, as well as Cancel, Clear and Go, which are used in response to questions posed by the instructor to cause the VCR to track to pre-programmed positions. The system offers a low-cost entry into interactive video, but the facilities are limited to simple branching and multiple-choice answers, with the questions separately provided on worksheets. The instructor is provided with a Cue Programmer to assist program production which places framing pulses and address information on an audio track of the videotape. Existing training materials can be 'Responderized'. A printed record of students' answers can be obtained, but no analysis of this data is done within the system. The Responder is currently only available to NTSC standard. A single student response unit costs some £300,

USING A RANDOM-ACCESS VIDEOTAPE RECORDER

and the system is designed to be attractive to those who already have appropriate VCRs and monitors.

Microcomputer/VCR interfaces

5.16 A number of purpose-built microcomputer/videocassette recorder interfaces are becoming available, of which a typical example is the CAVI 400 interface board developed by BCD Associated Inc and marketed in the UK by Michael Gurr Associates, Tenterden. This will connect an APPLE II, with 48K memory and a single disc-drive, to either a U-matic or industrial VHS player/recorder. The latter is to be preferred because of its faster search speed. Associated software is supplied which allows the videotape to be logged with framing pulses, and then the programme to be divided into numbered sequences that can be subsequently accessed by simple commands. An authoring system is also provided which will take a teacher step by step through the process necessary to specify the content of computer-generated 'pages', to set out multiple-choice questions, and to indicate the consequent steps to be taken in response to student answers. The authoring system assists both 'content generation' and 'control logic', but if the teacher wishes to exploit the graphics capability of the micro, or allow the student to manipulate a mathematical model, or undertake numeric or textual analysis of a student's answer, then the appropriate computer programs must be specially written.

5.17 A prototype system comprising microcomputer, VCR, interface and monitor can be assembled for some £3000.

5.18 Bryce (1982) has described a design for an interface that allows the switches on a Sony RX-353 Auto Search Control box to be accessed from a Cromemco System 3 microcomputer, by which means a U-Matic videocassette recorder can be programmed to search for different segments of material. A subroutine, written in BASIC, allows the videorecorder to be accessed randomly. Associated equipment offers the option of alternatively presenting computer-generated text or video pictures on a television monitor.

INTERACTIVE VIDEO

5.19 Other examples of interactive videotape systems which have been described as being available on the US market include the following.

Video-dex (Videodetics Inc), a self-contained programmer/controller/response unit compatible with most ½in industrial videocassette machines, which operates on a similar principle to the Responder, and is priced at about $1100.

Cavri Interface (Cavri Systems), capable of linking an APPLE II microcomputer to a Panasonic Omnivision II videocassette player and colour monitor to offer trainers self-programming facilities: total hardware package cost about $5500.

Video Mentor (Video Education Inc) is a microprocessor interface system controlling a videocassette machine and generating text materials, charts, questions and comments through an interactive authoring system, selling at approximately $4500.

CATI (Whitney Educational Services) will interface an Apple II computer to a videocassette player.

Bell & Howell have also demonstrated a modified Apple II interfaced with a Panasonic VCR for interactive instructional purposes.

6. OPTICAL RECORDING MEDIA

6.1 Advances in laser technology have made possible two related developments exploiting optical recording techniques: the videodisc and the optical digital data disc. Whereas interest in the videodisc is centred on its applications in the consumer market as a means of publication and distribution of television programmes and feature films, the digital disc is evolving from the interest of the computer industry in its potential as a mass storage medium in the information-handling field. Both devices have significant features that make them attractive in the educational and training arenas, although at this present moment more progress appears to have been made in bringing about the practical realization of the videodisc as a workaday tool than has been the case with the digital data disc. The common characteristics of the two devices are that they offer:

— compact, low-cost storage
— rapid retrieval of information
— environmental stability and archival quality
— portability.

Both systems are capable of storing text, numeric data, graphics and pictorial images, in monochrome and in colour, the fundamental difference being that the optical data disc records information in digital format, while the videodisc operates in analogue mode. A typical 30cm disc has a storage capacity of some 16 Gigabits in digital format, whereas the analogue videodisc can hold 54,000 separately addressable standard television frames per side.

6.2 Data storage capacity at these densities has obvious practical advantages, but these are matched by prospective economic factors that place optical recording in a distinctly separate domain from existing storage media. Cost comparisons with current media are shown in Table 6.1.

INTERACTIVE VIDEO

replicated by Discovision. Initial difficulties over reliability of the equipment and poor quality control in disc-mastering plant hindered market penetration, but considerable interest was aroused. Discovision launched a more rugged consumer player, followed by 'industrial' versions capable of degrees of interactive control. In mid-1980, Pioneer announced its own version of a consumer player for the USA market. All these players are compatible to the extent that they conform to the same disc format, but they each contain specific control and communication features that make total interchangeability between systems difficult. A cross-licensing agreement with Sony has led to yet another 'industrial' version of the player, and the establishment of mastering plant in Japan. Within the last few months Pioneer has taken over Discovision Associates, and now controls the California mastering plant in addition to its own two facilities in Japan. Pioneer is understood to be withdrawing the current Discovision players from the market in favour of its own new designs.

6.14 Philips launched a consumer version of their videodisc system in the UK in May 1982, and have built a disc-mastering and replication plant at Blackburn. This has been closely followed by the Pioneer company, whose new player again shows different features to that of Philips. These issues are taken up later in this report.

6.15 The emergence of the 'LaserVision' videodisc system has thus been long and complex, with several false starts and a history of apparent corporate uncertainty. The scale of investment required into both research and development, and into production plant, has been very high and to a certain extent has determined attitudes to market objectives and to the timing of developments. These have also been affected by the unforeseen rapid expansion of the videocassette market and the advent of personal computers. LaserVision is no doubt a technically sophisticated system which is capable of far more than replicating television and cinema films in the home, and despite the persistent thrust to harness the consumer field it is likely to be in the business and information dissemination fields that it can make most impact. This appears to be a view gathering acceptance within the manufacturing consortium, particularly with Sony, but it is difficult to obtain any clear statement of corporate strategies in this respect.

INTERACTIVE VIDEO

Optical transmissive systems have been produced separately by the French Thomson-CSF, and by the Atlantic Richfield Development Corporation.

6.8 The Thomson-CSF system relies on a transparent 'floppy' disc with the video information encoded in micropits on the surface. The disc is read by projecting the laser beam through the disc on to a photodiode. Since the laser can be precisely focused it can read micropits on both sides of the disc without turning it over. To avoid damage to the disc from dirt, grease and scratches it has a protective sleeve only removed when the disc is inserted into the player. Equipment has been in limited production for several years and NTSC, SECAM and PAL versions have been released. An 'institutional' version of the player was announced in 1979 containing a microprocessor allowing either use of a remote control keypad or connexion to an external microcomputer to effect control and programme sequencing. A Thomson-CSF TTV 3620 player interfaced to a Commodore PET computer has been used at the University of Berne for self-instructional medical programmes (Hufschmid, 1979). A number of commercial users in the USA have also experimented with the system for industrial training purposes.

6.9 Thomson-CSF has agreements with TEAC in Japan and with 3M in the USA to manufacture discs; the 3M mastering and duplication plant is in St Paul. Thomson is also reputed to have an arrangement with the Xerox Corporation to develop optical digital storage systems for business purposes. Following the recent nationalization of Thomson the future for these ventures is unclear.

6.10 The Atlantic Richfield subsidiary ARDEV has also worked on transmissive videodiscs but, in contrast to micropit coding, utilizes a photographic grey scale to represent the video signals. Diazo copies of the disc are read by an ordinary light beam, dispensing with the need for a laser. Advantages claimed for the photographic disc are that it uses a known and reliable technology and that mastering and disc

OPTICAL RECORDING MEDIA

production can be relatively cheap. One interesting technical feature is that each video frame can be associated with up to nine tracks of audio or digital information. Using audio compression techniques, this will offer up to 30 seconds of sound associated with each still-frame picture.

6.11 The ARDEV player has only been seen in prototype, and with its built-in microprocessor, keyboard and sophisticated software is almost a combined home computer-videodisc system capable of being linked with other computers and peripherals via IEEE 4888 and RS 232 interfaces. It is understood that McDonnell Douglas has taken over the future development of this system, with the aim of producing a player costing about $2500 and equipment that can be used for local recording at about $9000.

LaserVision

6.12 The practicality of an optical reflective videodisc recording system was first publicly demonstrated by N V Philips at Eindhoven in September 1972. By that time the American MCA (Universal Studios) had already set up a subsidiary, MCA Discovision, to develop opto-electronic recording techniques, and by 1975 agreement had been reached between Philips and MCA to amalgamate standards and jointly exploit the technology in the consumer entertainment field. Philips were to make the player, to be distributed through its Magnavox outlets, while MCA would be concerned with mastering and replication of software from its extensive film library. MCA then entered into an agreement with the Japanese Pioneer Corporation to build its own player aimed at the industrial/educational markets, to be followed by the creation of Discovision Associates as a 50/50 joint venture between MCA and IBM. This company maintained the arrangements with Philips and Pioneer to exploit the early patents, and also had interests in a number of special-purpose applications including document and image storage, digital read-only memories and talking encyclopedia. It set up disc-mastering plant in California.

6.13 Philips began marketing players in the USA in December 1978 with feature film software from Universal Studios

INTERACTIVE VIDEO

	Magnetic hard disc	Magnetic computer tape	Magnetic videotape	Silver halide	Optical disc
10 Gigabits equivalent storage	80 disc packs	90 8-track tapes each 2400ft	One 2400ft roll of 2in tape	200 fiche 4 x 6in	One 30 cm disc
Media cost	$40,000	$1350	$100	$60	$10

Table 6.1. Cost comparison of storage media

6.3 A brief survey of the prospects for the digital data disc as a component of computerized information-processing systems is given in Appendix 1. The bulk of this report is devoted to consideration of the characteristics of the optical videodisc, and its particular applications in the educational and training fields.

Videodisc technology

6.4 Videodisc technology has been developing since the late 1960s. The British Decca Co, working in collaboration with Telefunken, demonstrated an early system wherein video signals were recorded in a spiral groove on a vinyl disc in hill-and-dale method. The disc, similar to a long-playing audio record, rotated at 1500 rpm giving one full frame per revolution, and the signals were retrieved via a tracking stylus. Although marketed for some time in Europe it was never successful due to mechanical and wear problems, and was subsequently withdrawn.

6.5 Other systems began to emerge at about the same time. The Radio Corporation of America developed a more successful mechanical system in conjunction with the Zenith Corporation. 'Selectavision' also uses a grooved disc with a diamond stylus riding in the groove and retrieving the signal by detecting changes in capacitance. The system is only suitable for the consumer market and for viewing linear movie sequences, since it has no provision for still frame or

OPTICAL RECORDING MEDIA

variable speed facilities, and searching can only be carried out using an associated time index.

6.6 A second capacitance electronic disc (CED) system has been developed by the Japanese JVC Co in conjunction with the Matusushita Co. This design involves a flat stylus gliding over the surface of a smooth disc. The video signal is encoded in a spiral track within the disc as a series of pits which cause a difference in capacity between the stylus and the disc as the head tracks over it. In order to maintain the stylus in the correct position in relation to the spiral track a second set of pits encodes servo-tracking information to control accurately the head position. This VHD system records two full television frames per revolution, and since it is possible to skim the stylus across the disc both still-frame and rapid search facilities are available. There is still a residual problem of disc wear, however, and although a minimum disc life of 10,000 plays may appear reasonable, this equates to some few minutes only of playing time in still-frame mode. The advantages claimed are those of reliability and proven technology combined with relatively low cost. Thorn-EMI Ltd have an agreement to exploit the VHD system in Europe and have reportedly invested £20 million in disc production factories in Swindon and Cologne, with a capacity to produce 3 million discs a year. A recent announcement indicated that Thorn-EMI have now shelved indefinitely plans to launch the VHD system in the consumer field, as is also the case with their Japanese and American (General Electric) partners. Thorn-EMI, however, are still maintaining a small software development group investigating educational, reference and instructional uses of the medium.

Optical videodiscs

6.7 The development of optical recording and playback systems utilizing the high-resolution characteristics of laser light dates from about 1970. Two principles have been explored, one based on the modulation of a light beam as it passes through a medium, and the other based on the modulation of a reflected light beam from the surface of a medium.

7. THE LASERVISION SYSTEM

7.1 The LaserVision optical videodisc relies on the basic principle of a single information channel in which is stored all that is necessary to reproduce a colour television picture, together with two sound tracks and associated control data. In order to accommodate the non-linear characteristics of the chosen recording process the composite video FM signal is first modulated by a lower frequency audio carrier. This signal is subjected to amplitude limiting which results in a pulse-width modulation with two varying parameters: the length of the pulses and their distance apart. This process is illustrated in Figure 7.1.

7.2 This pulsating signal is used to modulate the intensity of a finely focused laser beam in the master recording equipment. This beams a micron-sized spot on the surface of a glass master disc coated with a light sensitive layer, producing a series of pits representing the original video signal. The master disc is then developed to a precise depth in a clean environment and a nickel stamper copy is prepared. This is used to replicate plastic copies which faithfully reflect in relief the pits on the master. Each plastic replica is coated with an aluminium layer to make it more reflective and a thin protective coat of clear plastic added. Two sides are glued together, making a rigid disc about 2.8mm thick.

7.3 Copy discs are read back in a player unit containing a low-power helium-neon laser which is focused through the protective layer on to the aluminized relief. The reflected beam is converted by a photodiode into an electrical signal that is demodulated and processed to regenerate an exact image of the original video signal ready for display on a normal television receiver. A diagrammatic representation of the recording and playback processes is shown in Figure 7.2.

THE LASERVISION SYSTEM

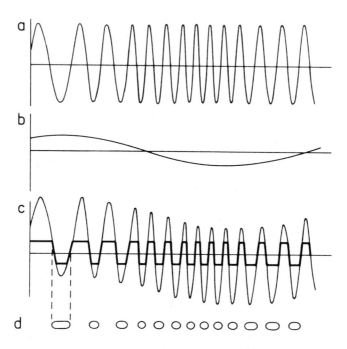

Figure 7.1. Modulation of video signal and subsequent limiting. (a) Video signal (b) Audio carrier (c) Amplitude limiting (d) Resultant 'analogue' pit pattern. (The Council thanks Philips Electronics for permission to use this material from their LaserVision manual.)

7.4   The standard LaserVision disc is 300mm in diameter. Information is recorded in a continuous spiral from the centre of the disc to the outside, which revolves anticlockwise as viewed by the laser beam. Due to the small size of the pits the disc can contain 54,000 'tracks' which can each be individually resolved. Since the laser is sharply focused on the internal reflective layer, it is insensitive to scratches, dust or grease marks on the surface of the disc, making it rugged to handle.

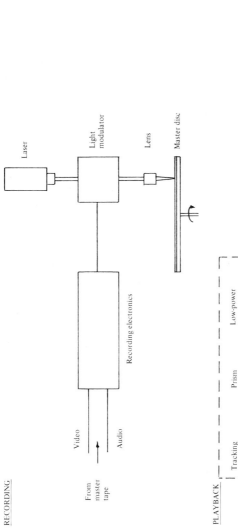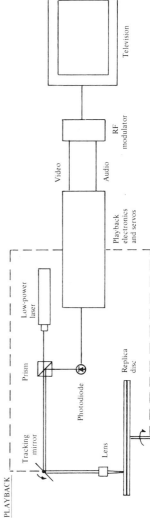

Figure 7.2. *Recording and playback processes — schematic*

THE LASERVISION SYSTEM

7.5 The disc can be operated in two modes. If operated at constant angular velocity (CAV) the recording is arranged so that one complete television frame (two fields) is placed on each track. Hence each revolution represents one video frame and the speed of rotation is synchronized with the standard television framing rate, 1500rpm for PAL and SECAM, and 1800rpm for NTSC systems. In this case the information is packed more tightly at the centre of the disc than at the outside. This arrangement makes it possible to achieve still-frame and slow-motion replay by reading the same track more than once, the optical beam jumping back during the vertical synchronization interval.

7.6 Discs recorded at constant linear velocity (CLV) are arranged with one television frame per revolution on the centre track, but with successively more thereafter. The rotation rate is automatically slowed down as the playing head moves towards the periphery. In this way twice as much information can be held on the disc, but at the expense of allowing no still-frame or slow-motion facilities. This is thus only suitable for recording long-playing programmes.

7.7 Coding information is also recorded in association with each full frame. In CAV recordings, each frame is given an absolute picture number, and it is also possible to insert chapter numbers to indicate the beginning of a new sequence. Such data is digitally encoded on lines within the vertical blanking interval, which do not normally carry picture information. It is also possible to code control information on these lines, such as stop codes, etc. CLV discs carry coding information that automatically disables the still, slow, fast and reverse player facilities. Instead of picture code CLV discs contain time code which allows the elapsed time of the programme to be displayed.

7.8 Both types of disc also contain vertical interval reference signals and vertical interval test signals. A minimum of 900 tracks prior to the start of the video programme contain a start code that positions the laser focusing head to the beginning of the programme at nine times its normal speed. A minimum of 600 tracks at the end of the

INTERACTIVE VIDEO

programme contain a code to direct the head back to the beginning at 75 times its normal speed. A comparison of the track layout of CAV and CLV discs is given in Figure 7.3.

Classification of LaserVision players

7.9 As indicated above, a number of different hardware manufacturers have designed playback units around the *de facto* standard for the LaserVision disc format. These each demonstrate different functional capabilities according to the degree of control, or 'interaction', they afford the user. R Daynes (1982) has suggested a classification of players in relation to their intended uses, which provides a crude identification of machinery with common characteristics. It is not to be expected, however, that equipment described as falling in the same group will be completely intercompatible, indeed this is the exception rather than the rule.

7.10 *Level 0.* This includes players with no inherent capacity to offer still frame, slow or reverse motion, or any real degree of addressability: for instance, the RCA Selectavision players, and those optical systems designated as 'extended play' or CLV solely designed to replay continuous, linear programmes.

7.11 *Level 1.* These are basically 'consumer' videodisc players having individual frame addressability and a worst-case frame access of better than 20 seconds. Such players will have limited memory and keypad control of functions, including still frame, slow, fast and reverse motion, but no real processing power. They may, or may not, have a connector to facilitate external control. Examples include the Philips VP 700 (PAL), Magnavox 8000 (NTSC), Pioneer VP 1000 (NTSC) and LD 1100 (PAL) machines.

7.12 *Level 2.* These are the designated 'industrial/educational' players specifically designed to be more rugged and offering all the facilities of Level 1 players, but with improved access times (worst case better than five seconds). These will have built-in two-way communications facilities to accommodate links with an external computer, and an onboard

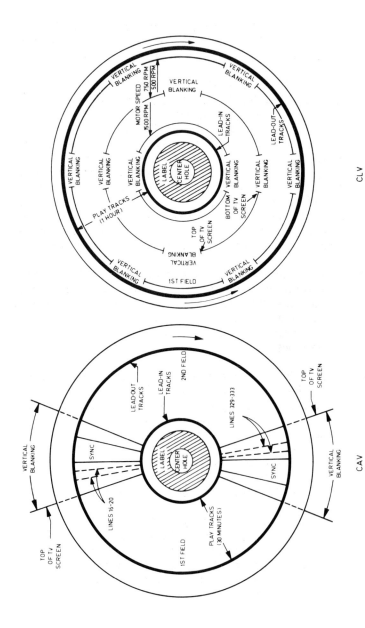

Figure 7.3. Layout of frame information on CAV and CLV discs. (The Council thanks Philips Electronics for permission to use this material from their LaserVision manual.)

INTERACTIVE VIDEO

microprocessor of limited capability offering a degree of 'intelligence' that can be pre-programmed either from the player keyboard, or by down-loading instructions directly from the videodisc. Examples of such players include the Discovision PR 7820 series (NTSC), Sony LDP 1000 (NTSC), and the prototype PAL player currently under development by Philips.

7.13 *Level 3.* This level categorizes equipment packages that include either a Level 1 or 2 player interfaced with an external microcomputer. These machines will offer facilities to interpose, or overlay, computer generated text and graphic images on pictures from the videodisc; full automatic control of all the player functions through computer programs; and the ability of the user to access all normal computer functions. No commercial systems yet appear to be on offer, although it is possible to construct Level 3 systems from various components, and interface units to interconnect standard personal microcomputers with videodisc players are becoming readily available.

7.14 *Level 4.* This level embraces more advanced systems in which the potential of integrated computer-videodisc hardware is exploited to the full. For instructional/educational applications such systems will incorporate comprehensive authoring languages and interrogative information databases to support a wide range of self-instructional strategies. Examples of such facilities include the computer-based learning system being developed by WICAT Inc.

Technical features of available hardware

7.15 Since LaserVision systems were initially developed for the American market, most experience to date has been obtained with hardware formated to the NTSC television standard operating with 525 lines at a framing rate of 30 cycles per second. Only very recently has any hardware conforming to the PAL standard of 625 lines and 25 frames per second become available, and this largely confined to the consumer versions. Much more data has therefore been published

THE LASERVISION SYSTEM

about the performance of NTSC players and this information predominates in this report. Where possible, however, specific data pertaining to PAL systems is given.

7.16 A comparison of the functional capabilities of various videodisc players is given in Tables 7.1 and 7.2.

	Magnavox 8000	Pioneer VP 1000	Sony LDP 1000	Discovision PR 7820 1/2/3
Still frame	*	*	*	*
Step frame	8	*	*	*
Slow motion	variable	variable	1/5	variable
Fast play	x3	x3	x3	
Direct frame access		*	*	*
Scan/search	*	*	*	*
Remote control		option	*	*
Two audio channels	*	*	*	*
Pause (CLV mode)	*	*		
Auto repeat	*	*		
Chapter stop	*	*		
Picture stop	option	*	*	*
Programmable			*	*
Memory (bytes)			1024	1024
External interface			RS 232	parallel
Microprocessor			Z 80	F 8
Digital disc dumps			*	*
Worst-case access (sec)	20+	20	5	2-5
Power consumption (W)	65	95	95	130
Weight (lb)	28	39	44	54

Table 7.1 Comparison of NTSC players

INTERACTIVE VIDEO

	Philips VP 700	Pioneer LD 1100
Still frame	*	*
Step frame	*	*
Slow motion	variable	variable
Fast play	x3	x3
Direct frame access	*	*
Scan/search	*	*
Remote control	*	*
Two audio channels	*	*
CX noise reduction		*
Pause (CLV mode)	*	*
Chapter stop	*	*
Picture stop	*	*
Programmable		*
Computer-control interface		*
Digital disc dumps		*
Worst-case access (sec)	25	10
Power consumption (W)	50	60
Weight (lb)	?	36

Table 7.2 Comparison of PAL players

7.17 No technical specification has yet been issued concerning the Philips prototype 'industrial' PAL player. Models seen are based on the LD 1000 player, but contain a microprocessor, at least 16K of memory and a teletext encoder. The prototype is also equipped with an external control bus in accordance with the RS 232C standard.

8. DESCRIPTIONS OF TYPICAL SYSTEMS

Level 1 systems

8.1 The Pioneer VP 1000 (NTSC) and LD 1100 (PAL) players, although designed for the consumer entertainment market, have features that make it possible to utilize them for a degree of interactive programming, and offer the possibility of linkage to a external microcomputer. The players can read and identify the various control codes and data that are placed, during the mastering process, in the vertical blanking intervals that occur between the fields of picture information. It can also store signals from its own keypad that define picture numbers and various commands input by the user. Since every picture on a videodisc has a unique frame number the control system can search for any single frame and display it. The search time depends on the physical distance between frames, but for two frames within 1000 locations of one another the response time is less than five seconds.

8.2 Commands from the VP 1000 keyboard are transmitted as 10-bit pulse-code modulation, and such codes can also be synthesised by an external computer. The player recognizes 25 different commands, which are transmitted as a series of pulse bursts at a carrier frequency of 28 kHz. Each pulse burst consists of 10 cycles and is exactly 0.263 milliseconds long (Figure 8.1a). To indicate binary 0 the elapsed time between one burst and the next is 1.05 milliseconds; and the duration between bursts to indicate binary 1 is 2.1 milliseconds. A command consists of a 10-bit sequence, and therefore contains 11 bursts and 10 'spaces' (Figure 8.1b). The actual code sequences are given in Table 8.1, where it can be seen that the first three bits are in each case identical, as are the last two, and that the central five bits vary to specify the command.

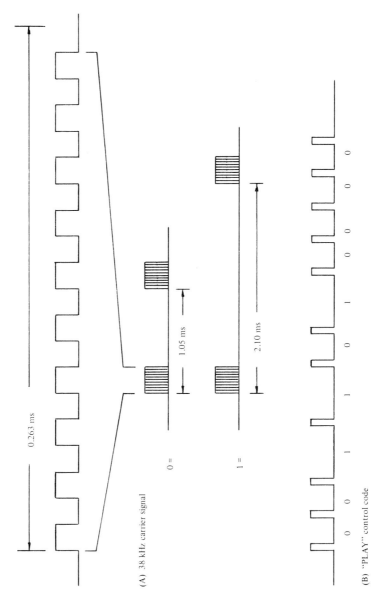

Figure 8.1 Pulse-code modulation of VP 1000 control signals

DESCRIPTION OF TYPICAL SYSTEMS

Function	Bit code value								
	K0	K1	K2	D0	D1	D2	D3	D4	D5	D6
0	0	0	1	0	0	0	0	1	0	0
1	0	0	1	1	0	0	0	1	0	0
2	0	0	1	0	1	0	0	1	0	0
3	0	0	1	1	1	0	0	1	0	0
4	0	0	1	0	0	1	0	1	0	0
5	0	0	1	1	0	1	0	1	0	0
6	0	0	1	0	1	1	0	1	0	0
7	0	0	1	1	1	1	0	1	0	0
8	0	0	1	0	0	0	1	1	0	0
9	0	0	1	1	0	0	1	1	0	0
Search	0	0	1	1	1	0	1	0	0	0
Chapter	0	0	1	0	0	1	1	0	0	0
Frame	0	0	1	0	1	0	1	1	0	0
Audio L/1	0	0	1	0	1	1	1	0	0	0
Audio R/2	0	0	1	1	0	1	1	0	0	0
Still/step										
Forward	0	0	1	0	0	1	0	0	0	0
Reverse	0	0	1	1	0	0	1	0	0	0
Scan										
Forward	0	0	1	0	1	0	0	0	0	0
Reverse	0	0	1	1	1	1	0	0	0	0
Fast										
Forward	0	0	1	1	0	0	0	0	0	0
Reverse	0	0	1	0	1	1	0	0	0	0
Slow										
Forward	0	0	1	1	1	0	0	0	0	0
Reverse	0	0	1	0	0	0	1	0	0	0
Mode										
Pause	0	0	1	0	1	0	1	0	0	0
Play	0	0	1	1	0	1	0	0	0	0

Table 8.1. Pioneer VP 1000 control codes

8.3 Circuit designs for suitable interfaces containing the necessary pulse-burst generator are given by Ciarcia (1982).

INTERACTIVE VIDEO

Pioneer offer a general-purpose interface at a cost of £80. A prototype interface specifically designed to connect a BBC microcomputer to a Pioneer videodisc player has been demonstrated by Orbis Computers Ltd, an Acorn subsidiary.

Level 2 systems

8.4 Industrial players to the PAL standard are not yet marketed, hence most published information relates to NTSC equipment. Two systems are available in the USA, the Discovision PR 7820 series and the more recently introduced Sony LDP 1000. The PR 7820/3 is the most recent update of the Discovision player, which has faster search times and an improved EPROM giving increased user-definable functions. Both players contain microprocessors and 1024 bytes of RAM, in which commands and registers can be stored. It is understood that the Discovision players are being phased out by their new owners, Pioneer.

8.5 The Discovision player contains a Fairchild F 8 microprocessor. This can be programmed to conduct a series of frame searches, stopping at prearranged points and waiting for input from the user, and then branching to the next sequence accordingly. Programs can be downloaded from the second audio track of the videodisc into the player's memory, or keyed in from the in-built keypad, or communicated from an external computer. The Discovision player has a non-standard parallel communication port, but an RS 232C interface (UEI) is available to connect with the serial port of the computer. The player can transmit the following information to the host computer:

— current frame number
— contents of RAM
— player status.

8.6 Each command signal requires one byte. Using the internal programming procedure, a sequence requiring the player to search out frame 2000, and then play to frame 5000 and automatically stop would be keyed in as follows:

DESCRIPTION OF TYPICAL SYSTEMS

Command	Byte	Comment
2	1	Enter start frame no.
0	2	
0	3	
0	4	
Search	5	Find frame 2000
5	6	Enter end frame no.
0	7	
0	8	
0	9	
Autostop	10	Play to frame 5000 and stop

8.7 By using registers, more economic use can be made of memory, which is particularly useful in a re-entrant program that stops and searches frequently. The above program might become:

Command	Step	Byte	Comment
1		1	
Recall			Activates register 1
2000			
Store			Stores 2000 in 1, activates 2
5000		2	
Store			Stores 5000 in 2
1	1	3	
Recall	2	4	Activates register 1
Search	3	5	Go to frame 2000, activate 2
Autostop	4	6	Play to frame 5000 and stop
End	5	7	

8.8 The Sony player incorporates a Zilog Z80 microprocessor, and uses a different programming structure. The functions of searching and autostopping are implied, and it is possible to define up to 63 different 'segments' of the videodisc to be called up in turn. This allows more efficient

53

programming, and additionally each command does not consume one memory location. The previous example becomes:

Command	Byte	Comment
Segment 1 = 1-2000	1	Define start and finish of
Segment 2 = 2000-5000	2	sequences
Stop	1	Stop at...
S001	2	Segment 1
Play	3	Play to the end of...
S002	4	Segment 2

8.9 More complex programs can readily be written using the internal logic of both machines to await input in answer to multiple-choice questions, and to branch accordingly backwards or forwards, and to make comprehensive use of the machine's still-frame, slow-motion and replay facilities.

Level 3 systems

8.10 Such systems are customized packages of videodisc player and personal microcomputer interfaced together to allow their joint operation. Since no agreed standards exist for either videodisc players or for microcomputers, a range of interface devices is appearing, some aimed exclusively for use with one player; some to accommodate the characteristics of several players; some exclusively linked to Apple II, Atari and other popular computers; and some generic interfaces suitable for most computers. Since most personal computers dispense with interlacing when generating video output, the sync pattern of the standard video signal as recorded on the videodisc is invariably incompatible with the sync pattern controlling the computer display. A simple solution is to switch the display between the two video sources, but a higher level of integration demands that both

DESCRIPTION OF TYPICAL SYSTEMS

outputs can be shown simultaneously, or overlaid. One approach is to modify the computer's video circuit to allow it to accept external synchronization pulses, or, more expensively, to digitize the computer video and transmit it via an interface board (with its own microprocessor) that will accept synchronization signals from the videodisc and regenerate the computer video.

8.11 To ease the process of communicating between computer and player, work has been undertaken to utilize high-level languages to program player commands, and to assist authors to construct instructional sequences at the computer keyboard. Bejar (1982) describes a series of UCSD PASCAL programs designed to control a Discovision player from a Terak microcomputer via a UEI interface. Other examples using BASIC and Apple PILOT have been mentioned.

8.12 Examples of complete systems and of specific interfaces are available on the American market (NTSC). Discmaster 5000, by New Media Graphics, combines a Pioneer VP 1000 with an Atari 400 microcomputer via a special interface with a built-in Z80 processor. The computer has 16K bytes of memory and a floppy disc-drive, and programming is done in BASIC. The Omniscan (Aurora Systems Inc) links the Pioneer VP 1000 to an Apple II computer by an interface that allows the Apple to simulate operation of the player's keypad, and automatically switches video output between player and computer. The software does not allow overlays. The VMI interface from Allen Communication will link the Apple II computer to all existing optical video players except the Magnavox 8000.

8.13 Despite widespread popularity of the Apple II microcomputer, it is not the most suitable machine on which to base an extensive videodisc operating system. It is probably nearing the end of its life, and it has the serious limitation that it is difficult to integrate its non-standard video output without an expensive additional interface board. There is more advantage, therefore, in seeking to use a new generation machine with a lot of development potential, and

INTERACTIVE VIDEO

here the new 16-bit machines stand in a very powerful position, and to seek a system that can support a comprehensive authoring language.

8.14 Special-purpose systems have been built by Positron and by WICAT Systems with an emphasis on simplifying the authoring of programmes and text and image generating without need of computer programming skills. Emerson Electric Co has designed an interactive video classroom to teach motor skills and offer simulation activities, combining the Sony videodisc player with a purpose-built controller unit. The student communicates with the system via a touch-sensitive screen over the monitor, thus eliminating the need for a keyboard.

8.15 Mention should be made of video-processing interfaces that permit mixing of computer output with standard television signals. Sanders Associates' ITVS will mix Apple II generated text and graphics with NTSC video and respond to digital data carried on the blanking lines of the video signal to download instructions. Video Associates Labs' Micro-Keyer will also reprocess Apple II output, and is particularly used to overlay weather maps and forecasts directly into local broadcasts.

8.16 The use of teletext technology to intermix computer output and video images does not appear to have made much ground in the USA; however, a number of interesting developments incorporating teletext with videodisc are emerging in Europe. These are more fully described in Section 9.

Level 4 systems
8.17 More advanced systems and combinations of media facilities involving videodisc equipment are very much in early stages of development. The best known work is that carried out in Professor Negroponte's group at Massachusetts Institute of Technology, where their Spatial Data Management System incorporates two videodisc players and a computer colour graphics processor under the control of an Interdata 32-bit computer. Output from the twin discs and the computer

DESCRIPTION OF TYPICAL SYSTEMS

generated imagery is mixed in a special video effects unit ensuring that there is no loss of continuity between sequences. Sound from any of four tracks can be played with a picture from either disc or sound-over-stills provided. The colour graphics processor can create overlays for interactive control menus, highlight elements of the video images, and display viewer annotations. A touch-sensitive screen and joystick controls are provided, and the system logic is such that the program is always responsive to the viewer, who can cause changes to the program sequence at any point rather than only at fixed 'branching nodes' (Backer, 1982). The system has been extended to offer the facilities of a complex flight simulator, with dynamic image processing, at a fraction of the cost.

8.18   An interesting development has taken place under the auspices of the American Heart Association, where a robotic dummy with built-in sensory devices has been linked to an interactive disc system to provide instruction in methods of resuscitation. The program not only illustrates correct procedures, but monitors and evaluates student performance (Hon, 1982).

9. VIDEOTEX AND VIDEODISC TECHNOLOGIES

9.1 Videotex is the generic term embracing broadcast teletext information and viewdata information services transmitted along telephone networks. Teletext services broadcast coded signals representing 'pages' of written or simple graphic information at the standard television transmission frequencies within the vertical blanking intervals of the picture frames. On a normal television set these signals are invisible, but with a special teletext decoder circuit the signals can be reconverted to their original form and displayed on the screen. Each page can contain 24 lines of written information, with each line capable of holding 40 characters. This is some 150-200 average words. Seven colours, including white, are available; and these colours can also be used to generate backgrounds and block images. The standard character set includes both upper- and lower-case characters, and figures; and can be displayed in double-height format.

9.2 Viewdata similarly allows the projection of pages of textual or graphic information on the standard television set, but this time the data source is held in computer memory and is transmitted via the ordinary telephone network. It is necessary to have a special decoder/adapter at the television set to convert the transmitted signals to standard television frames. As with teletext, each 'page' can display 24 lines of 40 characters. In the case of the UK PRESTEL service this is an 'alphamosaic' display, offering seven colours but only rudimentary graphics. Other systems allow better graphic resolution — for instance, the Open University OPTEL service will display 240 x 320 points against PRESTEL's 72 x 80 — and can operate in 'alphageometric' or 'alphaphotographic' modes.

9.3 Broadcast teletext services are limited in the number of pages that can be transmitted at one time unless a whole channel is

VIDEOTEX AND VIDEODISC TECHNOLOGIES

dedicated to the service. Pages are normally transmitted sequentially and captured by the viewer by keying in the appropriate page number on a keypad. It is possible to superimpose teletext on a normal television programme, as is done to obtain news flashes and subtitles. Circuitry allows part of a teletext page to be enlarged, sequences of pages viewed, or single pages held.

9.4 Viewdata offers a virtually unlimited source of information pages, determined solely by the capacity of the computers supporting the service. Utilizing 'gateway' techniques, privately owned computers can be connected to the public PRESTEL network, enabling subscribers to gain access to a whole new population of database sources. Access to certain databases can be restricted to closed groups of users, and it is equally possible to install a private 'in-house' viewdata system to offer a wide-ranging message-distribution and information-handling service for individual organizations. Viewdata pages are organized in a hierarchical tree structure, with each page uniquely numbered, accessed by keying in the appropriate page number, or by following a predetermined route indicated by the information provider. Viewdata services can offer limited two-way communication, by which users can respond to information providers via simple 'response frames' or, with more elaborate equipment, transmit electronic mail.

9.5 In many instances, microcomputers can be adapted to act as viewdata terminals. This is a particularly significant feature, since this offers opportunities to store PRESTEL pages in local memory, and to print these out in hard-copy form. PRESTEL can also be used to transmit computer programs, since these can readily be converted into viewdata format and automatically downloaded into the user's own microcomputer memory. CET has developed a transmission standard for such 'telesoftware', which has been accepted by British Telecom as a recommended protocol for program format. Such standards will hopefully shortly extend throughout Europe. A commercial telesoftware service on PRESTEL, Micronet 800, was launched in February 1983.

INTERACTIVE VIDEO

9.6 Broadcast teletext can also be used to carry computer programs. Special 'intelligent' television terminals are required to capture and decode the signals from the specially compiled master program and dump these into a cassette recorder. The BBC/Acorn microcomputer can also be adapted to receive programs broadcast over the BBC Ceefax service.

9.7 The interaction of microcomputer and television technology that has led to the current state-of-the-art of videotex (teletext and viewdata) technology has an obvious extension in relation to its integration with videodisc technology. Indeed, certain steps have already been taken in this direction. The relative ease with which videotex data can be mixed, and concurrently displayed, with camera-produced video offers operational and economic advantages despite the current limitations on the quality and resolution of videotex images. The introduction of 'second generation' viewdata services using high-speed integrated services digital networks, and enabling alphaphotographic (colour photographic quality) images to be received, will give an added boost; as will access to an increasing range of information programmes and sources via cable and satellite links.

9.8 As described earlier, the CAVIS interactive videotape system incorporates a videotex generator to enable keyboard-entered text and graphic symbols to be mixed with camera-generated recordings on the videotape. The prototype Philips 'industrial' videodisc player incorporates a teletext encoder, which allows a programme author to specify messages and questions to be interposed from computer store, or directly read from the videodisc. It is necessary to provide a television monitor with the requisite decoder. The 'active play' videodisc produced by BBC Enterprises for the LaserVision launch, David Attenborough's BBC *Videobook of British Garden Birds*, contains Ceefax encoded text on 210 pages to offer in-depth information on the 70 species shown in the programme, and which are displayed as still frames on command. Computer systems that provide the means of generating videotex will have particular significance as interface equipment in interactive video systems.

VIDEOTEX AND VIDEODISC TECHNOLOGIES

9.9 Videotex is not in itself a particularly attractive medium for the direct presentation of learning materials in programmed learning mode, due to the limitations of its format and branching structure. It has, however, marked attractions when seen more broadly as a resource, and as a means of communicating between students, teachers and institutions. Applications of PRESTEL-type services to education are currently being studied and several interesting pointers are beginning to emerge, with specific emphasis on private viewdata systems and the 'gateway' concept.

9.10 The Open University's Optel private viewdata system uses the same standards as the British Telecom public service and aims to improve communication to its dispersed tutorial staff, and eventually to students. It provides an improved character set and graphics, for mathematics, diagrams and maps, and offers a name-based indexing system with keyword retrieval. An information bank holds back-up information on current courses, and use is forseen in providing access to bibliographic reference material and research reports. It can also be used to circulate topical news. The 'electronic mail' facility of viewdata systems could become an increasingly important channel of communication to students, and a means by which students may readily contact their tutors.

9.11 Gateway services can offer similar opportunities for educational users to gain access to specialized services via the public network. Some of these services might be linked specifically to interactive video programmes, providing complementary facilities such as: encyclopedic databases; real-time handling of volatile information; organizational and administrative back-up; mailbox and message circulation. Information may be transmitted back from the student on the progress of his studies, or requesting tutorial help. It would be possible to implement a fully transactional service with virtually the same terminal equipment as needed for a stand-alone interactive video system.

10. SYSTEMS SOFTWARE

10.1 Software requirements for interactive videodisc systems can be divided into a number of discrete elements, distinguished by the purpose or function each serves.

Control programs

10.2 These are needed to identify the disc characteristics, ready the player mechanism, and ensure correct synchronization of the video carrier. These are embedded in the information on the disc itself, and are incorporated by the disc manufacturer during the mastering process. Although it is expected that all mastering facilities will adhere to a common specification, this is not absolutely certain, nor is it explicit that current practice will be followed for all time. This has implications for compatibility between discs produced at different mastering centres, and for archival material.

10.3 A 'start code' recorded in the lead-in tracks on the disc is used to trigger the playback mechanism and move the laser beam to the beginning of the programme. Conversely, an 'end code' appears in the lead-out tracks which mutes the video and audio signals, and which resets the scanning mechanism for the next play. If it is desired to incorporate chapter numbers or automatic still frame commands on the disc then cue codes pertaining to these must also be inserted, indicating:

— field dominance
— chapter number
— automatic still frame signal.

These codes are generated by the LaserVision Cue-code Inserter. In the PAL system, frame lines 14, 15, 19 and 327, 328, 332 are reserved for these codes, and may be used for nothing else. In addition Vertical Interval Test Signals

SYSTEMS SOFTWARE

can be incorporated in lines 16, 17, 18 and 329, 330, 331, to assist compilation. However these signals will be blanked out during mastering. Lines 20, 21 and 333, 334 are reserved for teletext data.

10.4 If these cue codes are being recorded on 1in type C helical videotape format, then it is essential for the VTR to have a fitted sync option to ensure satisfactory recording. Timecode, which has to be continuous and sync-locked, is also recorded during the mastering process; and video reference 'colourbars' and 'video black' and audio 'pilot tone' are required during the lead-in period. Video black is again required during the lead-out period. These codes are recorded as 24-bit biphase signals, and will be accessed by the player.

Command codes

10.5 These are the signals acted on by the player that organize the sequence of functions it follows during the course of a programme. These commands may be generated by keypad input, or be pre-recorded on the videodisc (in an audio track) and downloaded into the player's memory, or be generated from an externally linked computer. Examples of key codes used by the Pioneer VP 1000 and LD 1100 players have already been indicated in Table 7.3. 10-bit pulse code modulated command signals are created by depressing the keys on the player's remote-control keyboard. Pioneer have also demonstrated a computer interface which connects to any RS 232 computer port. Command signals generated by the computer are accepted by the interface and transmitted via an infra-red LED link to the player's remote sensor using the same pulse code as before. The command code used by Pioneer is given in Table 10.1. In addition to the commands equivalent to the functions offered from the player keyboard, there are four additional commands to facilitate computer control. Since there is no handshaking from the player, the video output is continuously monitored to indicate whether a previous command (eg, Search) has been fully executed. The 'video wait' command causes the next step in the computer program to be delayed until

Function	Computer code
0	0
1	1
2	2
3	3
4	4
5	5
6	6
7	7
8	8
9	9
Search	S
Chapter	C
Frame	F
Audio 1/L	L
Audio 2/R	R
Still/Step Forward	O
Reverse	I
Scan Forward	M
Reverse	N
Fast Forward	H
Reverse	G
Slow Forward	K
Reverse	J
Mode Pause	Q
Play	P
Computer	U
Videodisc	D
Video Wait	W
Reject	E

Table 10.1. Control codes for Pioneer computer/videoplayer interface

the presence of a video signal again indicates that this is so. Further, since no frame count information is returned by the player, frame counts must be simulated by

SYSTEMS SOFTWARE

counting timing pulses within the computer to enable pause commands to be transmitted at the correct moment. Using such an interface fairly simple interactive programs can be created using a high-level language such as BASIC. The Orbis prototype interface uses similar techniques, but more neatly, and aims to employ the NPL Microtext authoring language.

10.6 A provisional description of the computer interface for use with the Philips prototype industrial player has also been issued, and is reproduced in Appendix 2. This has a 45-element command structure, as shown in Table 10.2.

10.7 The Philips industrial player is designed to communicate with an external computer via an RS 232 C bus, and can offer 'handshaking' to indicate the completion of control actions. Given the information outlined in Appendix 2 it is not difficult to construct a suitable high-level language implementation of the command structure to simplify program creation. In addition, up to 800 pages of teletext information can be generated from the external computer keyboard and stored.

10.8 It is clear that different equipment manufacturers are taking separate approaches to the design of command protocols, a process that could become aggravated as a range of interfaces appears designed to link with different models of microcomputer. Some agreement on common protocols at an early stage would appear to be in the best interests of users concerned to maximize interchange of courseware.

Authoring software

10.9 This is required to assist subject experts and materials designers to create learning programmes without extensive knowledge of computer programming or system design. An authoring language, such as PILOT, will provide a framework within which a programme producer may create frames of textual information, set questions for the student and indicate the various actions to be taken dependent on the student's responses. Many computer-based learning systems

HEX	DEC	CHARACTER	COMMAND	TIME TO EXECUTE (ms)
2E	46	.	Correction	
2F	47	/	Enter	
30	48	0	0	
31	49	1	1	
32	50	2	2	
33	51	3	3	
34	52	4	4	
35	53	5	5	
36	54	6	6	
37	55	7	7	
38	56	8	8	
39	57	9	9	
3A	58	:	Pause	
3B	59	;	Memory	
3C	60	<	Search Reverse	
3D	61	=	Auto-stop	
3E	62	>	Search Forward	
3F	63	?	Picture Number	
40	64	@		
41	65	A	Audio 1 ON/OFF	200
42	66	B	Audio 2 ON/OFF	200
43	67	C	Chapter Number	200
44	68	D	Picture Number	200

45	Video Mute ON/OFF	40
46		
47		
48		
49		
4A		
4B		
4C		
4D		
4E		
4F		
50		
51		
52		
53		
54		
55		
56		
57		
58		
59		
5A		
69	E	
70	F	
71	G	
72	H	Remote Ext. ON/OFF	40
73	I	Keyboard Active ON/OFF	40
74	J	Remote to Int. ON/OFF	40
75	K	Key Release	40
76	L	Still Forward	100
77	M	Still Reverse	100
78	N	Normal Play Forward	200
79	O	Normal Play Reverse	200
80	P	Go to Picture No.	
81	Q		
82	R	Run	40
83	S	Slow Speed Change	40
84	T	Teletext	60
85	U	Slow Motion Forward	200
86	V	Slow Motion Reverse	200
87	W	Fast	40
88	X	Clear	40
89	Y	Repeat	40
90	Z	Plus	40

Table 10.2 Command table for prototype Philips industrial player

INTERACTIVE VIDEO

offer authoring languages with these facilities, but most suffer from the limitation that both logic and content are inextricably bound up in the same computer program.

10.10 Greater flexibility and range of options is afforded by a new generation of authoring systems that allow instructional logic and instructional content to be kept separate. One authoring system designed specifically with interactive video in mind is VCIS developed at the University of Utah, which consists of a series of PASCAL routines. Another is the WISE system developed and marketed by WICAT Systems.

10.11 WISE operates on the 'menu' principle, whereby the author is presented with a range of choices to define the pattern that the lesson material should take and, having made a selection, the system takes charge of the programming detail, seeking from the author just the necessary data for the displays, and criteria for judgement and routing. WISE allows decisions on branching to be made by the author (predictive), the student (free interaction), or by the system (conditional). Student responses are monitored and assessed against different sets of criteria, such as multiple-choice selection, spelling, word order, correctness, keyword inclusion, synonym matching, arithmetic accuracy, number of tries, etc. The system will offer feedback for right and wrong answers, and links to other lesson materials, or to PASCAL subroutines custom-designed to deal with simulations or other specific points. Displays may consist of computer-generated text, with or without high-resolution graphics, or animation. A 'help' display can be called up at any time. Finally, the system will collect and analyse statistical data about student performance.

Student-oriented software
10.12 This may be required to allow students to access the computer in calculation mode, and to allow annotation of displays and generation of graphical inputs in response to prompting. Students may also need to interact with simulation programs, or use the keyboard to construct free responses in answer to questions. Responses may be by

SYSTEMS SOFTWARE

touch-sensitive screens, light pens, scratch pads or cursor-controlling joysticks. Such software should be totally transparent as far as the student is concerned, particularly so where materials are aimed at less-able or disabled students.

11. DATABASES FOR INTERACTIVE VIDEO

Video images

11.1 The videodisc is designed to hold 54,000 full frames of video per side, which offers some 37 minutes of playing time in PAL format. (The NTSC format offers 30 minutes per side). In interactive mode it is unlikely that continuous viewing of a programme that long will be required, and the expectation is that the disc will be broken into a series of shorter segments, or 'chapters', which can be each individually addressed. These segments will normally be produced by standard studio methods, and may well consist of extracts from existing footage if this is of sufficiently suitable quality to meet the mastering specification. There is a wealth of existing material that could lend itself as primary sources for interactive video databases, much of it of a documentary or historical nature, some of which will have been specially produced for educational and instructional purposes. The bulk of this archive will be in the hands of the broadcast television networks, and commercial film companies and libraries. This will create serious problems of copyright clearance, and require negotiation of secondary rights of use. It is to be hoped that the owners will look sympathetically at requests to have access to such materials.

11.2 It is clear, however, that to take best advantage of the new medium it will, in nearly every case, be necessary to create new material, as introductions to, or as inserts into, existing footage; and in many instances, to produce completely fresh programmes. The design philosophy behind such new programming must adapt itself to certain radical concepts:

— programme elements may not be viewed continuously, or necessarily consecutively
— programme elements will be used and re-used by the same viewer, and must be free of irritating features

DATABASES FOR INTERACTIVE VIDEO

— there is an expectation that certain elements will require to be updated on a regular basis
— it is not necessary, or even desirable, to regard the video medium as the sole carrier of the message.

11.3 Within the category of video databases must be included the use of animation techniques, and dynamic graphic devices, to convey educational ideas and information. Such materials may be produced by traditional rostrum camera techniques, but can increasingly be expected to be generated with the aid of sophisticated computer graphics facilities.

11.4 Methods for the preparation of materials for transfer from video or film media to videodisc are discussed in Section 13.

Still images

11.5 Still images can be recorded as single frames on a videodisc and individually accessed. Provided necessary precautions are taken to ensure correct field interlacing, still pictures can be edited from videotape recordings, but it is more likely that still images will be transferred by teleciné techniques from photographic slide or film copy. Still images can be real-life photographs, or copies of specially prepared artwork, or computer generated graphics; and may be in monochrome or full colour. Animation sequences are in effect series of consecutive still images.

11.6 The theoretical capacity of the videodisc of 54,000 images per side makes it a powerful and compact medium for storing visual databases, and some interesting examples are being proposed. The use of the videodisc as a component in a visual information retrieval system has considerable potential, as would be the production of 'resource' discs, which are not in themselves programmed but which offer the the prospect of being used in conjunction with locally generated computer programs. Discs containing purely single image frames will conceivably be used in a learning context, but a major application of the single frame capability will be to insert individual frames between video image segments to supplement and amplify the dynamic information. The placing and editing of such material will be a significant feature of programme design, particularly if the search characteristics of the player units are to be used to best advantage.

11.7 If the user is to be offered the opportunity to select his own pictures from an extensive database of still images, some assistance with indexing, or other search procedures will need to be provided. An experimental still-image database has been produced by MIT, and the NASA Jet Propulsion Laboratory is proposing to place 100,000 space probe pictures of planetary surfaces on disc; other examples include application to accessing hospital X-ray films and pathology microscope slides, geographic maps and charts, and architectural and engineering design data. The major limitation at present is the relatively poor resolution of the television screen, which causes fine detail to blur or disappear altogether. An optimum resolution figure to support high-quality graphics is about 1250 lines, which is twice the line standard of current broadcast video.

Textual information

11.8 Pages of text can be stored on videodisc by considering them as equivalent to still images. They can thus be photographed directly from printed or graphic artwork, or copied from books and documents. There are limitations to the type and quantity of print that can be accommodated on a standard television screen. Some fonts and typefaces are more easily perceived than others, and the optimum capacity of a frame is no more that 30 lines of 64 characters each, or less than 2000 characters. Moreover, the aspect ratio of the standard television screen is completely different from that of the common document formats, and a conventional A4 typewritten document would normally require eight television pages to offer satisfactory reproduction.

11.9 Text can also be stored on disc in videotex format by creating 'pages' from a teletext or viewdata keyboard. Teletext (Ceefax) messages can be encoded on frame lines

DATABASES FOR INTERACTIVE VIDEO

in the vertical blanking intervals, and viewdata (PRESTEL) signals recorded as modulated analogue in an audio channel for subsequent decoding. However, with systems linked to a microcomputer, it is probably most convenient and flexible to store textual material in digital format on floppy disc and call this up for presentation by program subroutines. This technique offers a range of options, dependent on the type of VDU available and the characteristics of the interface. Text can be stored in teletext format, to be interposed between or overlaid on video images; or can be standard computer alphanumerics in a variety of fonts; or can be mixed with computer-generated graphics, etc.

Computer data files

11.10 Program code and associated data files are also best held on floppy disc, or similar backing store, although it is again possible to encode computer programs on videodisc. The risk of drop-out during the mastering process, though not significantly serious for analogue video signals, can produce irretrievable errors in digital code. The development of reliable optical digital storage discs will, however, offer possibilities of enormous and fast, but cheap, read-only memories that will transform the capacity of microcomputers and challenge the alternative technologies of hard-disc or networking.

Audio databases

11.11 Audio signals associated with video pictures are recorded in real-time in the same signal channel on the videodisc, and decoded on playback to offer two separate channels — these can be a stereo pair, or two distinctly different soundtracks. It is also possible to use audio compression techniques to allow longer bursts of audio to accompany picture presentation, a facility particularly valuable with still-frame presentation. Audio can be digitally recorded on magnetic tape or disc, but the storage requirements for a high-fidelity digital recording are severe.

11.12 For voice outputs, pulse-code modulation can be used as a simple digitization technique; an even more economic technique is to model the human vocal tract and store only that data necessary to regenerate approximately the same sound — the technique used by Texas Instruments in their talking games. Rather than modifying and storing input from a human voice, it is possible to directly synthesise voice output from stored representations of speech sounds (phonemes and allophones) and so create audio messages by typing in the string of speech sounds.

Integration

11.13 A hardware system comprising a videodisc player connected to a microcomputer with associated floppy disc memory is thus, potentially, a very powerful and versatile information storage and retrieval unit. Such a self-contained system can provide all a user needs to access databases containing moving and still full colour pictures, textual and other alphanumeric files, audio messages, and all the necessary computer code and operating software to make the system function. Additional databases can be provided by a library of videodiscs and floppy discs, or by linking the system through a communications network to distant service points.

12. COURSEWARE CONSIDERATIONS

12.1 The educational value of an interactive video system is dependent, above all else, on the quality of the content materials and the manner in which these are made available to the would-be learner through the working out of the overall pedagogical design. It is perhaps inevitable in the early stages of development of any new technological delivery system that much emphasis will be placed on technical detail and on the opportunities offered, and limitations set, by the new medium. Successful acceptance of interactive video systems will, however, be largely influenced by the subtlety and discrimination with which learning experiences — the courseware — are presented, and on the inherent nature of the informational content, which must not only be of an appropriate level and style but also able to attract and hold the attention of the student.

12.2 The powerful combination of microcomputer and videodisc offers a unique range and variety of presentation and control over teaching materials for the individual learner, but this very versatility makes it all the more crucial to build its applications around the best practices that have already evolved within the constituent educational technologies. It is clear, however, that strategies hitherto thought the epitome of educational television or computer-based learning practice will need considerable adaptation.

12.3 Moreover, despite the attraction of interactive video as the ultimate, stand-alone, individual learning environment, it is likely to prove most effective when used complementarily with established presentational media, and not seen as the sole channel for communication between teachers and taught.

INTERACTIVE VIDEO

Presentational features

12.4 The videodisc is primarily an audio-visual medium, offering simultaneously dynamic visual and audio modalities. The availability of two audio tracks offers the possibility of alternative commentaries to the same visual material; for example, in different languages, or at different levels. Video material can be broken down into discrete segments, covering specific pieces of information, or teaching points, which may be accessed randomly, or in any desired sequence. Additionally, such sequences may have locally generated information added to them, through overlaying computer-generated text, or incorporating separately recorded sound.

12.5 Audio messages can be presented with a totally blank video screen, or accompanied by a still image taken from the disc or generated by computer. Or still images can be presented alone. The image storage capacity of the videodisc, and the possibilities of local magnetic storage, offer virtually limitless combinations of single-frame sequences.

12.6 The computational capacity of the system can be exploited in simulation exercises, where decisions made by the student may be instantly incorporated in sequential presentations of data. Likewise, the students may be offered the use of the computational facilities to work out their own solutions to problems.

12.7 The informational capacity of the system can be extended by linking the local microcomputer to distant databases via telecommunications links, of value where rarely required information or archival material is of consequence; or in instances where access to rapidly changing data is essential.

12.8 The limitations of the video screen for presenting large quantities of textual or tabular information should not be forgotten, and it is essential that in all those instances where reference to conventional printed materials is more advantageous, workbooks and associated supplementary reference sheets should be provided, even if locally printed from electronically stored sources.

COURSEWARE CONSIDERATIONS

Instructional strategies

12.9 When considering the design of interactive video programmes the Nebraska Videodisc Group divides material to be incorporated into a number of broadly defined sets (Daynes, 1982):

orientation: frame sequences that set out the contents, objectives, prerequisites and other background information needed by the learner to identify the lesson material
subject content: these are the main instructional sequences, containing all the necessary illustrative and explanatory material required by the learner — as well as the main pathway, alternative sequences may be provided to offer remedial instruction, different pace or style of presentation, or recall of earlier work, etc
decision: frames that require some input from the student to determine the next course of action
comment: frames giving advice or assistance to the learner on how best to progress through the material to achieve the learning objectives
summary: concluding sequences that summarize the content of the lesson, or fragment
test: problems to test the learner's mastery of the subject matter so far presented, and worked examples that offer the learner the option of self-evaluation.
help: a set of frames that offer assistance to students who become lost or confused during the course of a programme.

12.10 While such a classification is superficially helpful it must not be viewed as too categoric, and indeed it would be feasible to define information frames and sequences in many other ways. For example, there will often be a place for purely reference data — ie, there is no instructional message given or implied — which may be indexed or cross-referenced for easy access by the student. Such reference material could well be common to a number of programmes, at various levels of difficulty.

Programme design

12.11 As a general observation the use of any new communications medium evolves in a pragmatic way as innovators experiment with the new form and recipients comment on and criticize their artefacts. Gradually an accepted corpus of wisdom emerges, encapsulating the collective experience (in terms of styles, approaches, techniques), described in terms of a correspondingly evolved critical apparatus, which to all intents and purposes becomes a *de facto* theory. Although there is perhaps a satisfactory sense of understanding of what makes an acceptable and effective novel, piece of music, work or art or film, there is less agreement on the use of the same media — the printed word, sound, still and moving pictures — when these are applied to the purposes of teaching and learning. Given that even the textbook, with several hundred years of evolution, remains a limited educational tool, what considerations should determine the manner in which the elements of interactive video are best applied to educational ends?

12.12 O'Neal and Lipson (1982) have given some thought to this problem and suggest it is important to distinguish between acquisition of declarative knowledge, or 'knowing what', and procedural knowledge, or 'knowing how'. The value of computer aided learning has been largely in its ability to facilitate 'knowing what', and although its interactive capability is significant the computer can normally only accept limited responses and offer fairly rudimentary guidance on the basis of such responses. To improve the capability of CAI systems, and by extension interactive video systems, current computer intelligence needs augmenting to include aspects of human intelligence — possibly through development of heuristic 'expert systems' — and by more effective use of visual stimuli.

12.13 Whereas declarative knowledge can be effectively taught by words and symbols alone, the teaching of procedural skills often requires parallel sound and visual images. Further, the ability of the human brain to perceive and recall visual images has as yet been but poorly exploited in teaching practice. The potential of computer/videodisc systems to offer models of skilled performance, and to simulate complex situations in both work-oriented and interpersonal areas offers a major challenge to curriculum designers. To this may be added the need to ensure that greater attention is given to specifying student activities that enhance and confirm 'knowing how' rather than just testing students on content.

12.14 A further dimension offered by the video medium is that it can readily provide an emotional element to assist attention and focus learning. The selective use of images chosen for their beauty, drama, surprise, novelty or challenge, is perhaps of equal importance to issues of format, colour, typography, and other physical attributes.

12.15 Finally, O'Neal and Lipson recommend discriminating use of the various channels of communication offered by the new technology, on the one hand to avoid confusing the learner with a plethora of conflicting stimuli, but on the other to take greatest advantage of the human ability to abstract the most significant information from the environment.

Learning styles

12.16 It is perhaps on the issue of learning styles that interactive video raises the most interesting questions to which, at this point in time, there are no distinct answers. Previous educational technologies have usually been very limited in their accommodation of different learning styles, often forcing an assumption that all learners approach a particular learning task in the same way. Evidence from comparative examination of alternative techniques often fails to recognize that for part of each population the most efficient method is not the one being tested. Interactive video offers the possibility that a range of different learning styles may be accommodated within the single overall system, and therefore that attention may be given to the holist as well as to the sequential learner; that conceptual development may precede or follow practical experience; that provision can be made for a student wishing to review material part-forgotten or misunderstood, as well as for the first-time candidate; that allows the experienced learner to be highly selective, but ensures that the naive learner does not get by with only a shallow skim, and that both have opportunities to manipulate and explore the learning environment to their own best advantage.

12.17 Since there is no clear prescription, educators will have to learn to apply interactive video 'on the run', and in so doing should understand a lot more about the psychology of learning. This is one reason for a coordinated programme of research in this field that can both feed off and contribute to the further development of the technology.

13. VIDEODISC PRODUCTION

13.1 The design and production of an interactive videodisc has many similarities to the scripting and production of a conventional film or television programme, except that there are several further considerations. These arise from the need to produce appropriate computer programs to match the visual sequences; to anticipate the range of interaction between the user and the medium; to take account of the technical constraints imposed by the disc mastering process and to undertake sufficient evaluation before the final, irrevocable, mastering since no further editing is then possible.

13.2 The elementary components of a videodisc programme can come from many sources — motion film, videotape, photographic slides, audio recordings and computer generated displays. Such source material may be selected from existing stock or may be specially commissioned original material, but in either case care has to be taken to ensure the highest possible quality copy is acquired and that it has not already suffered degradation. The process of producing videodiscs involves several stages of signal transfer, and if any of the contributory material being collected for the mastering process is already more than three generations distant from the 'original' there are likely to be noticeable and irritating blemishes in the final discs.

13.3 Since the disc production process is totally in the hands of the commercial mastering agencies, the key objective as far as an educational producer is concerned is the achievement of an adequate 'pre-master' copy. The 'pre-master' can either be a completely edited 16mm or 35mm film, or an edited and frame-synchronized videotape in either 2in Quad or 1in type C Helical format. Videotape is largely the preferred format: the mastering specification issued by Philips for

LaserVision is reproduced by permission in Appendix 3. Technical issues to consider when assembling material for mastering are discussed below.

Film

13.4 Standard film framing rates are 24 per second. When transferring to the PAL television system the common practice is to ignore framing rate differences and to speed up film transport to match the 25 frames per second of the television signal. This four per cent difference is largely unnoticeable, but may be critical in certain sequences, or where sound pitch is crucial. When transferring to the NTSC system, teleciné equipment automatically arranges for successive film frames to be scanned twice and then three times to accommodate the 30 frames per second television signal. Additionally, the aspect ratios of film and television frames are different, which may give rise to distortion in one dimension, or loss of information. Particular care has to be taken when it is intended that images from motion film sequences should be viewed in still frame. Unless both fields of the interlaced television picture are matched to a single film frame, unacceptable flicker will result.

Slide

13.5 Standard 35mm slides can be used as source material, but again it is important to remember that the television aspect ratio differs from that of the film format, and that there is a limited 'safe picture area'. Television resolution is also less than that obtainable on film and thus fine detail may well be lost. It is particularly necessary when photographing text, that fonts should avoid serifs and slanted lines, to avoid 'staircase effects' on the television screen, and to remember that the maximum display capacity for easy legibility is 25 lines each of 64 characters.

13.6 The incorporation of slides and related still-frame material into videodisc programmes is not a trivial matter. Cost estimates for preparing a photographic slide and transferring it correctly and with appropriate coding to a pre-master videotape range from £6 to £60 dependent on the complexity

VIDEODISC PRODUCTION

of the original image. Given that a complete videodisc could contain 54,000 such single frames per side, this implies a massive investment at the production stage. The choice and number of single frames incorporated thus should be carefully assessed.

Videotape

13.7 Videotape sources for videodisc productions should be in 2in Quad or 1in type C Helical format and preferably be original tapes. Editing videotape sequences, and incorporating material from other formats on the final pre-master tape can only be satisfactorily done on computer-controlled editing equipment to ensure accurate frame synchronization. The same equipment may be used for adding and mixing sound tracks. The availability of a broadcast-quality videotape editing suite is a prerequisite for videodisc production, and if this facility is to be sought from facilities houses costs of the order of £300 per hour can be expected. A typical rate for on-line editing in a carefully pre-planned session might be 40 edits per hour.

Electronic image handling

13.8 Wherever practical, the use of electronic image generation and handling techniques should be considered, since these offer potential savings in effort and cost in preparing material for interactive video applications. Animation produced from graphic board artwork is notoriously expensive, as is any manual creation of drawings and text. Computer generated graphics and text in a form that is compatible with camera-generated images allows for a more flexible and integrated approach to creating and editing an educational programme, and access to sophisticated computer graphics and caption generation equipment can facilitate overlay, image merging and text insertion tasks. Given access to a videotape library, it would be possible to edit out under computer control single-frame images at far less cost than obtains when working from a slide collection. The value of wordprocessing techniques when developing scripts and text goes without saying.

Authoring technique

13.9 The design and production of an interactive video programme is a highly complex affair demanding the integration of a number of professional skills. It is a process that requires considerable planning and organizational management if it is not to be wasteful and unproductive. At root must be a fairly precise identification of the target audience, which, although ultimately meant to be the individual working alone, will have in most cases to be a member of a more widely defined group in order to ensure economic viability. It is worth spending a significant amount of time at the analytic stage, since once production has begun the costs of false trials can mount disastrously.

13.10 As with a conventional educational television programme the starting concept is expanded into a draft script and storyboard, but from the earliest stages a flowchart becomes imperative as the deviations from the straight linear sequence of normal programming make their appearance. As with programmed learning and computer-aided instruction a systematic application of the precepts and practices of instructional design should guide the evolution of the programme, but is essential not to lose the overall view (not to lose sight of the wood for the trees) and to become fascinated by the mechanics of the delivery system at the expense of the message.

13.11 Since the videodisc has evolved out of conventional television technology, it is inevitable that much emphasis will be placed on the predominant role of the television producer. However, this predominance is not necessarily so and should soon become eclipsed as was the early dominance of computer specialists in the development of computer-aided learning. In interactive video both sets of skills and professional experience are desirable, as is also the critical subject matter expertise and a strong measure of pedagogical drive and motivation. The conclusion is that the more successful enterprises are likely to be team efforts, based on the model of the Open University, or of the US Science

VIDEODISC PRODUCTION

Curriculum Improvement Study, capable of producing high-quality material at a high level of output and at reasonable cost. It is unlikely that much success will be achieved by lone individuals operating as 'cottage industries'.

13.12 The authoring process can thus be expected to flourish best in an environment with considerable logistic support, and where concentrated effort over an extended timescale can be brought to bear on the task.

13.13 One variant on the concept of complete authoring by a specialist design team that is felt worth exploring would be to evolve a series of basic interactive video programme components that could be welded into a specific course by a local teacher or trainer with the aid of a do-it-yourself authoring system. This would reflect the traditional practice of the teacher selecting from various professionally produced textbooks and other learning materials that which most suited his particular students. The ability to create, with the aid of a personal microcomputer, individualized programmes in this manner would appeal to certain teachers, although it is not foreseen that this should become the predominant methodology.

Editing and evaluation

13.14 The compilation of an interactive videodisc programme is thus a complex process with the final editing stage being the key activity. At this stage a number of requirements converge — the need to work to demanding technical standards; need to ensure completeness and overall effectiveness of the learning sequences; and the most economic ordering of the material to minimize editing time and to optimize disc search times when actually in use. It is recommended that a draft edit off-line is accomplished, which should highlight any particular difficulties or discrepancies in the programme. This edit can also be used for formative evaluation of the programme, since there is no opportunity to insert revisions once the mastering process is initiated.

13.15 Where possible, a simulation facility should be available, based on a random-access videotape player. At the simplest level this can be a videocassette recorder under microcomputer control, able to play back sequences in the anticipated order, to check accuracy and logic, etc. Preferably a broadcast-quality VTR, with frame-accurate editing facilities using computer-controlled time code should be available for final editing and evaluation, since this will allow revisions to be made *in situ* and apart from longer search times offers all the functions expected of the videodisc.

13.16 When all revisions have been made to the author's satisfaction, the complete pre-master tape is then edited up on-line, fully synchronized, and with time code added in one pass, taking into account all the detailed requirements of the particular mastering facility to be utilized. At this point the pre-master is handed over to the mastering facility, who will insert the disc coding signals and frame numbers, and produce a master tape from which the master videodisc will be produced.

Mastering

13.17 Even with the standard LaserVision system there are minor variations in the actual mastering process, dependent on the manufacturer employed — Philips, Pioneer (ex-Discovision), Sony or 3M. In a typical process an optically ground and polished glass plate is covered with photoresist to an accurately controlled depth. After curing, this is exposed in the recording equipment to the modulated laser beam, and developed to produce the characteristic pitted surface. Following a procedure similar to that used for audio long-playing records, a nickel stamper is made, from which can be replicated the customer discs. Customer discs are formed by thermoplastic injection moulding, with the replicate surface first plated with an aluminium reflective layer, then covered with a protective clear plastic coating, and two sides bonded together to make a double-sided disc.

13.18 A trial disc is usually offered for review purposes, although this may only be rejected for major mastering errors or excessive drop-out. The reliability of the manufacturing process appears to have improved considerably since initial

VIDEODISC PRODUCTION

services were offered by Discovision Associates although it is not possible to guarantee total error-free coding of digital, frame numbering data. Drop-out effects on the video signal are usually not so troublesome unless extensive. For certain applications, to ensure that all frames of a programme can be retrieved, these are recorded in duplicate. The introduction of immediate read-after-write mastering equipment should offer better guarantees of effective mastering.

13.19 Once replicate discs have been received and checked, the final touches can be made to accompanying computer programs, ensuring that these are completely debugged and that all frame references agree with the master coding.

Production costs

13.20 The costs of mastering discs are individually assessed and dependent on the extent of the preliminary work the mastering facility has to undertake. However, given a satisfactory pre-master copy a typical charge for mastering would be £2000-£2300, with an additional charge of about £10 for each single-sided replicate disc produced. The cost of replicate discs is highly dependent on the numbers involved, and would be correspondingly more for very short runs.

13.21 The average time taken for a mastering facility to produce customer discs to special order appears to vary from 6-12 weeks.

13.22 There is no simple rule-of-thumb that can be used to compute the total costs of a typical interactive video programme, since this depends wholly on the complexity of the production, the range and type of material it comprises, and the extent to which existing or original footage is to be incorporated. Various figures have been hazarded ranging from about £30,000 to convert a largely extant video production to simple interactive videodisc format, to suggestions of £50,000-£100,000 per programme where the production requires extensive original material

INTERACTIVE VIDEO

and a corresponding amount of computer programming and text authoring. Figures of up to £250,000 could be reached for productions consisting largely of single still-frame compilations. Since all such figures are based on extremely limited, and largely American, experimental activities, it is clearly very difficult to be categoric about overall costs, and further developmental work needs to be carefully monitored to assess long-term cost implications.

14. RESOURCE REQUIREMENTS

14.1 The resources that are likely to be called upon for the design and production of an interactive video programme have already been touched on in various sections of this report. These considerations are now brought together and further commented upon.

Manpower skills

14.2 It is suggested that most progress will be made if interactive video development is considered a team activity. No specific recommendation is made as to which professional group should take the prime responsibility for leadership, since this may best vary from project to project, but the following areas of skill and expertise can be readily identified as contributing to a viable operation:

— subject expert and authoring skills
— teaching and curriculum development skills
— instructional design and evaluation skills
— text editing and graphic design skills
— computer programming and system engineering skills
— film and television production skills
— project direction and budgetary management skills.

14.3 It is not essential that all these manpower resources should be coexistent; however, there should be ready means of communication between contributing members to a project, and committed access to their services and the facilities they command. For the optimal conduct of each project it is suggested that the core of leading members, composing the management team, should be no more than six people.

14.4 The complexity of the processes involved in interactive video production, and the interrelating nature of the contributing professions will require that extensive and up-to-date documentation is maintained on each project. In particular, steps should be taken to identify general purpose routines and to identify and index fundamental teaching materials, and those documentary and archival images that could have multiple application. Projects will therefore need significant secretarial and logistic support, with specialist information search and library services.

14.5 An important aspect of programme compilation will relate to the use of existing material subject to copyright and performing right restrictions. There is obviously an exceptional quantity of existing film, television and still photographic material that could be adapted for interactive use, some of which will be unique material which cannot possibly be re-shot. It is to be hoped that the owners of existing rights will not prove unnecessarily obstructive in making such items available. In any case, production teams will need the services of experienced negotiators to secure clearances, and to ensure their own rights are not violated.

Equipment and facilities
Television facilities

14.6 The major requirement is that of access to broadcast standard television production facilities. This is not to say that acceptable interactive video programmes cannot be assembled by groups without origination capacity in the video field, and it is perfectly possible painstakingly to piece together filmed materials and still photographs without recourse to television studios or cameras. However, there are many advantages in electronic image handling and video editing; and the final process has in any case to be conversion to an acceptable, high-quality video signal, so that eventually the services of a well-equipped educational television unit or facilities house are inevitably needed.

14.7 To create a comprehensive interactive video production facility *ab initio* would be unjustifiably expensive, and therefore it is sensible to look to existing facilities that might form the basis of such a service. An obvious example

RESOURCE REQUIREMENTS

is the recently completed BBC/Open University Television Production Centre at Milton Keynes, which has already been involved in experimental videodisc production. Other educational television facilities exist at universities (including London, Leeds, Cardiff), and polytechnics (Brighton, Middlesex, Plymouth, etc), or are operated by local education authorities (ILEA, Clwyd, Chiltern Consortium, etc); or others like the late Road Transport and Distributive Industry Training Boards' centres have been involved in industrial training applications of television. Almost without exception, these units comprise adequate studio areas, but tend to be underprovided with modern broadcast standard equipment, and are currently woefully understaffed and underfinanced. They represent considerable public investment. The spare capacity of some, at least, of these centres could, it is suggested, be gainfully utilized by judicious upgrading of equipment and increasing technical support establishments.

14.8 The types of equipment most likely to be lacking in these educational television units include:

— flying-spot teleciné scanners   £50,000-80,000
— time-code synchronized videotape editing
  suites   £35,000-75,000
— PAL/NTSC converters   £30,000
— ENG cameras   £4,000-7,000
— computer-controlled graphics/text
  generators   £35,000-50,000
— digital image processors   £17,000-25,000
— digital slide stores   £25,000
— post-production mixers   £35,000-45,000

14.9 Although it is possible to consider the use of facilities houses for aspects of post-production work, and final pre-master editing, this is likely to be an expensive option in the long run. *Ad hoc* use of facilities houses (or freelance crews) is probably unsatisfactory in production work, where close collaboration with teachers and authors will be required, and denies the build-up of pools of expertise in the specific practice of educational programme making. It is recommended, however, that tight control of shooting schedules and management of studio facilities should be exercised as on commercial lines.

Computing facilities

14.10 It is less easy to specify the computing requirements for the design and production of interactive video programmes since, unlike the video component, there is less standard practice with which to identify. There will be a temptation to consider the use of simple, limited microcomputer systems for the authoring and frame composition process on the basis that these are the most likely machines to be in the hands of the users. However, this could prove to be a false economy, both in the time taken to create any appreciable quantity of courseware, and in terms of the range of facilities open to the author.

14.11 It is felt that any computing system for production should be sufficiently comprehensive and powerful to offer a subject expert or teacher/author, who will not necessarily be a computer expert, a very ready and straightforward vehicle for their creative talents. It should be capable of receiving a range of inputs, from keyboard, datapad or joystick, etc, and be able to access on-line stores containing relevant databanks, program libraries and other utilities. It should have very powerful text-processing and editing facilities, and be capable of generating good resolution graphics in colour in a form compatible with the standard interlaced television raster format. It should be interfaced with a videodisc simulator based on a frame-accurate editing VTR, and be provided with a frame store, and multiple monitor displays, for rapid comparison of material. It should have a high-quality text and graphics printer, suitably buffered to allow new processing to continue while output is being printed.

14.12 The system should be capable of supporting several high-level languages, including those commonly used, and should have an extensive authoring system with separate logic and

RESOURCE REQUIREMENTS

content addressing. The authoring system should allow design or specification of any type of display (frame), and the separate designation of the decision structure that will determine the subsequent display. In order to reach appropriate decisions the system may analyse student inputs in specified ways, update and compare stored statistics, or seek further input from the student. It should also allow the calling of high-level or machine-code routines. It should provide an intelligible 'help' and 'diagnostic' service, and will with advantage be menu-driven.

14.13 Developmental computer systems should be available to interactive video producers based on or compatible with the microcomputer machinery commonly available in schools and colleges, particularly Apple II, RML 380Z, and BBC/Acorn. The two former machines are, however, based on outdated technology, and at least one system based on 16-bit chip technology (for example WICAT Systems M68000, which has an impressive authoring system) should be included. A suitable WICAT system would cost £20,000-25,000, and whereas a system based on Apple, 380Z or BBC micro would cost far less in hardware terms, several man-years' effort will have to be expended to develop a comprehensive authoring system. In effect, the total cost would be similar. The developmental work begun by Acorn/Orbis should, nevertheless, be encouraged.

14.14 Separate computer systems will need to be established for software and system development, particularly for the development of comprehensive authoring environments, but also for the development of interface and control systems, extension of input and output facilities to accommodate audio communication, touch sensitive screens, scratchpads, etc. As such advances are made they may be transferred to the production systems. The engineering of systems to operate on a variety of user terminals should be an important aspect of the development programme. In this respect software design should embrace modular principles and aim, as far as is possible, to be machine independent. Applications programs should be capable of recognizing the

INTERACTIVE VIDEO

physical system they are to be used on and adapt to the specific hardware characteristics. This will require the establishment of common interface specifications and protocols.

User terminals

14.15 The basic user terminal is expected to consist of a videodisc player unit interfaced to a general-purpose microcomputer. This is thought to be preferable to introducing custom-built logic into player units. At this point in time the only PAL-compatible players generally available are domestic units with a limited range of functions. These commonly market at about £500 each. There is a strong likelihood that 'industrial' player units, with more extensive capabilities, will emerge on the market by the end of 1983, at a cost of about £1000.

14.16 A typical general-purpose microcomputer and interface to work in conjunction with a videodisc unit will probably cost about £500, although a more comprehensive system, including monitor unit disc-drive and printer would cost a minimum of £1500. It is therefore likely that the initial target cost for user terminals can be placed at £2500, with the expectation that this price might reduce to £2000 within 18-24 months, if a sufficient market materializes.

14.17 The effect of the rapidly expanding market in cheap home computers (under £200), currently estimated to have reached 500,000 units in the UK, should also be considered in relation to the development of interactive video. It has not been possible to explore the full implications of this phenomenon during this study, but it could indicate a significant potential market for simple interactive programmes in the recreational education field. The characteristics of the popular personal microcomputers should therefore be taken into account during system development work.

15. ECONOMIC CONSIDERATIONS

15.1 The exploration of the potential of interactive video techniques for educational and training purposes and the development of the first examples of applications programmes will require significant investment. This fact has already been recognized in the USA where a number of research programmes have been initiated with assistance from federal and institutional funds. Among these are the Videodisc Design/Production Group funded by the Office of Science and Technology of the Corporation for Public Broadcasting at the University of Nebraska's television station KUON-TV; an evaluation project covering 45 school systems at the American Institute for Research in the Behavioural Sciences, funded by the Department of Education through the Division of Educational Technology of the Office of Library and Learning Technologies; a project at Dartmouth College funded by the Sloan Foundation; and a project at the National Center of the American Heart Foundation. Advanced work is also being supported by the Defense Advanced Research Projects Agency at MIT. There are many other examples of appreciable institutional and commercial investment in interactive video development in the USA.

15.2 The scale and range of resources necessary to make any real progress is such that special funding arrangements are imperative, at least during the initial exploratory and developmental stages. It is natural to look to government, or government agencies, in the first instance to provide financial support. The question is to what extent could such support arguably be justified, and whether the traditional mechanisms of public funding of technological development projects are appropriate in this instance.

15.3 There are different aspects of investigation into the application of interactive video techniques to education and training that could implicate various arms of government and related public bodies. For example, the Science and Engineering Research Council might be persuaded to take an interest in the computer engineering aspects of interfacing microcomputers and television technologies. The Social Science Research Council through its Educational Research Board might accept a concern to support psychological or evaluative studies associated with use of the new medium. The Medical Research Council/DHSS could have an interest in its applications to the initial and continuation training of the medical and para-medical professions; the Manpower Services Commission would perhaps see its applications as a powerful instructional system of value in the development of the Open Tech programme, or more generally useful in the industrial and vocational training fields; the Department of Education and Science and the local education authorities would be concerned to assess its implications on the future provision of educational services, and the British Library has already indicated an interest in the applications of videodisc technology to information dissemination; the BBC, through its responsibilities for educational broadcasting, (and by implication the IBA) would wish to determine its impact on the future style of educational television.

15.4 There is potentially, therefore, a plethora of public interests involved, not least those of the Council for Educational Technology with its remit to advise on the application of all technological innovations of value to education. Likewise the Department of Industry, the co-sponsor of this study which has a twofold interest in encouraging educational practices of value to industry, and of encouraging the development of new, 'knowledge-based' industries which might offset our national decline in manufacturing potential.

15.5 There is a danger that any piecemeal approach to research and development in the field of interactive video will be of limited effectiveness, and possibly wasteful of resources, since it is unlikely that any one agency would feel it proper, or be able, to supply the level of funding required to

ECONOMIC CONSIDERATIONS

follow through all the interrelated aspects of the field. The relative isolation of separately funded groups could lead to duplication of basic work, and slow overall progress. A coordinated approach to the organization of a national programme of research and development for interactive video is thus recommended.

Scale of programme

15.6 From consideration of the resource requirements outlined above it is recommended that a research and development programme is initiated as detailed in Section 16. This should run in the first instance for three years, and will broadly require a capital investment of £3 million with revenue costs totalling some £5.75 million over that period. It is proposed that only part of this funding should be provided directly from the public purse, and that significant industrial and commercial involvement is sought. There are four reasons for this attitude. One is that interactive video techniques can be equally applied to general educational activities and to industrial and commercial training requirements — the fundamental technological equipment and processes are expected to be similar, if not exactly the same. Secondly, a closer collaborative relationship between sectors of industry and the formal educational system should be welcomed — and particularly when each partner stands to gain advantage from the product to be developed. Thirdly, the programme should be definitely applications-oriented, and would therefore benefit from business management techniques and concern with budgetary control and timescales expected of a commercial venture. Lastly, but most importantly, the programme should be seen as leading to the creation of a viable new industrial partnership capable of producing and marketing educational and training systems in national and international markets.

15.7 The proposition is that an interactive video development consortium be established in which government and industry are equal partners. Half of the necessary funding would come from government (or from appropriate budgets of contributory government agencies), the other half would come from industrial and commercial sources. For example, in this instance a possible division of interests in percentage terms might be:

Government interests	50
Electronics industry	10
Computer industry	10
Publishing and distribution industry	10
Television and communications industry	10
Merchant banks	10

15.8 The consortium, which might be considered a pilot for a more widespread educational development agency, would be responsible for developing a collaborative research and development programme and channelling funds to projects. It would organize shared facilities and retain exploitation rights in all products of the collaboration. The industrial partners of the consortium would have first refusal in the commercial exploitation of the products, and royalties accruing to the consortium would be partly used to fund further activities and partly to repay the investing partners. In this way the risks involved in launching a substantial enterprise would be considerably shared. It is suggested that the government should not attempt to recoup its own original investment, and that it should be largely responsible for the capital element, presuming this to be substantially invested in equipment retained in the public sector. A prime object would be to reach a position by the end of the third year whereby decisions on the future viability of the 'industry' could more accurately be made.

15.9 When identifying the scale of activity to be recommended, in which there has to be an element of speculation, an alternative approach was sought to that of building up an estimate from individual components. An analogy was taken from the field of commercial television production, where figures for the separately identifiable costs for an extensive series of programmes was available. The Channel 4 television series 'Brookside' has a specifically identified annual budget of £2.3 million, and a capital investment in equipment of

RESOURCE REQUIREMENTS

some £1 million. For this abut 50 hours of television transmission per year is to be produced — an hourly rate of about £50,000.

15.10 If we equate the effort required to produce one hour of broadcast television with that required to produce an interactive video programme comprising a one-sided videodisc and associated computer software, then this too would cost on the average £50,000 per programme. (Such a programme might occupy a student for some 8-12 hours — equivalent to one week's work in a subject.) This is put forward as a not too far-fetched comparison. If we now postulate that the objective of the interactive video development is to achieve an output of 50 programmes a year in the third year of operation, the intermediate costs can be scaled, making assumptions about the rate of expansion of output.

15.11 Further detail relating to the proposed nature of the research and development programme is given in the next section.

16. A MODEL RESEARCH AND DEVELOPMENT PROGRAMME

16.1 It is proposed that a collaborative research and development programme should be established with the following objectives:

— to identify and develop the technical means (hardware and software) by which interactive video can be used in an educational and training environment
— to establish well-equipped facilities for the design and production of interactive video programmes
— to create a series of teaching and instructional programmes to be used to attempt a critical evaluation (including cost-effectiveness) of the technique
— to establish the basis of a new 'knowledge industry' exploiting interactive video
— to set up a training and information dissemination service for prospective authors and producers, and for teacher users
— to set up a representative forum of educational and training interests, which a particular concern for specifications and standards.

16.2 The programme should provide for the establishment of up to four interactive video production facilities. These should be based on existing educational television facilities, a number of which should be invited to bid for consideration. These facilities should already be substantially endowed with high-quality equipment and basic support staff, and able to demonstrate a high standard of production. The centres would be obliged to give a commitment to allocate adequate studio time and resources to support at least 10 interactive video productions each by the third year of the programme. It is expected, however, that the selected centres will require additional or upgraded equipment, and additional funding for staff and running costs. It is estimated that the average

A MODEL R & D PROGRAMME

additional capital investment per centre will amount to £200,000 in the first year, followed by amounts of £100,000 in the two subsequent years. Additional revenue costs will be largely governed by the assumed production schedule, and will rise from an average of £100,000 per centre in the first year to £500,000 per centre in the third year. The centres might each specialize in distinct application areas, and it would be sensible to concentrate, rather than disperse, specialist facilities.

16.3 A fifth centre should be designated as a training and information dissemination centre, which should be equipped to the same general standard as the main production centres, but which will largely be utilized for training production staff, authors and programme designers, and user teachers. It will also be available for occasional experimental production work.

16.4 As well as providing computer systems at the production centres, a number of remote authoring units should be allowed for, complete with a random-access videotape recorder to simulate videodisc operation. These units would offer subject experts in other locations the opportunity to prepare, off-line, learning materials for later incorporation into video productions. It will also be necessary at the beginning to establish computer facilities to support the development of authoring systems and systems control software. At least one advanced microcomputer facility should be established. Unless access is available to a comprehensive computer graphics generation system at one of the production centres, this may need to be separately provided. The computer systems should be linked by telecommunications lines, in order to share facilities. Capital provision of computing equipment is estimated to total £300,000 over the three years, with annual support costs of £150,000 per annum.

16.5 At the beginning of the second year 200 user terminals, each costing £2500, should be purchased for distribution to cooperating user institutions. During the third year a further

INTERACTIVE VIDEO

400 terminals should be purchased, but this time it is assumed that the price will be trimmed to £2000 each, and that the users will contribute half the cost. An amount should be set aside for terminal maintenance during the third year.

16.6 The revenue budgets of the production centres are estimated according to the number of productions that would be scheduled for completion. The assumptions are that 10 productions would be completed in the first year, 30 in the second and 50 during the third. The production budgets are to include the employment of authors and all contributory consultants, if these are not to lie where they fall. In certain instances, it will be preferable to commission part or the whole of a production exercise from a commercial facilities house, or industrial training consultancy. No separate provision is made for this eventuality and it would be for the project direction to determine how these costs would be covered within the overall budgets.

16.7 An assumption is made that of the first year's productions, none would be truly cost-effective because of their experimental nature. Of the second year's productions it is assumed that half would be cost-effective in their own right (though not necessarily saleable) and that of these perhaps five might be exploited in wider markets. Of the final year's productions at least 30 should be cost effective, of which 15 should be marketable more widely. Exploitation of these materials should thus begin to generate income for the consortium, in those and subsequent years. As use of the medium develops, users would be expected to pay for all services offered by the consortium.

16.8 Within the budget of the training and information centre, provision should be made for the servicing of a representative user forum, to act as a general point of exchange of information, to focus issues of concern between users and manufacturers and suppliers of equipment, and to coordinate work on specifications and standards. Until it is possible to launch an integrated research and development programme, some means should be sought of establishing an

A MODEL R & D PROGRAMME

interim user forum, possibly under the auspices of CET, to deal with urgent user issues and maintain a watching brief on developments in the UK and abroad.

17. APPLICATIONS OF INTERACTIVE VIDEO IN EDUCATION AND TRAINING

17.1 The potential applications of interactive video technology in the educational and training fields are virtually unlimited, and any attempted listing becomes either a dreary catalogue or an arbitrary selection of curriculum areas. Interactive video will prove an attractive extension in those areas where computer-based learning methods have already made an impact, and offers opportunities for introducing computer-based techniques in topics where hitherto inability to deal satisfactorily with images and sound has been a limiting factor. Although general areas of application can be readily identified, the success and impact of specific interactive video programmes will depend strongly on detailed analysis of user requirements, the effect of traditional attitudes to technical innovation, and the financial parameters of the situation. Fortunately, television is already an accepted and largely respected medium in the educational world, and the rapid spread of interest in microcomputer technology suggests that the climate for the introduction of interactive video is promising.

Schools

17.2 The acceptance of new educational technologies by schools has been slow and patchy, from a combination of general lack of understanding amongst teachers, institutional inertia and critical cost thresholds. With the spread of videodisc into the domestic market it is expected that equipment prices will fall to levels affordable by schools, that publishers will begin to distribute materials, some of which will have peripheral (if not direct) interest to schools, and that many teachers will gain passing familiarity with the technology. Encouragement to seek alternative methods will come from pressures to reduce the amount of directly broadcast educational television during school hours; and from pressures to provide wider and wider ranges of educational opportunity with reducing teacher resources. The possibilities

APPLICATIONS IN EDUCATION AND TRAINING

of employing systems that allow greater reliance on 'showing how' rather than 'telling about' could prove attractive in overcoming endemic classroom problems of motivation, underachievement and relevance. A further point is that the use of information technology in schools will mirror the use of that same technology at work and in the home, and by so familiarizing young people with the technology we help them develop the appropriate skills to make use of it.

17.3   In the primary sector the main focus is likely to be on the basic skills of reading and number, although there will be applications in the elementary stages of foreign language learning and other elements of the primary curriculum. There are opportunities of exploring the value of pictorial and audio sequences to assist concept formation, and to introduce young children to a wide range of real world and social experiences. There is a potential market of some 28,000 primary and nursery schools in the UK.

17.4   In the secondary sector (over 6000 maintained and independent schools in the UK) the multi-media features of interactive video make it attractive in subject areas such as history, geography, art and drama, where pictures, photographs and moving sequences can be very illuminating. It also has potential applications in the social sciences and the humanities, and in the acquisition of life and communication skills. Of particular applicability might be a set of 'resource centre' compilations, which could be used by teachers to illustrate features in class lessons, but which would be equally a component of individual study programmes taken under computer control. Typical examples might include compilations on the history of art, on the development of science and technology, and of contemporary archives to encourage social and political awareness.

17.5   In the sciences, the videodisc offers a new dimension to the employment of computer-aided learning methods in that it opens up the possibility of more realistic simulations of the practical aspects of the work, which are often becoming increasingly difficult to organize and fund in schools. It

INTERACTIVE VIDEO

also offers opportunities to introduce pupils to wider aspects of engineering and technology, not hitherto regular features in school curricula. A related sphere is the area of vocational education, where practical experience is an important component, again not always readily available within the classroom. The demonstration of procedural skills, whether of shop or office practice, or in using workshop equipment is possible. In many instances there is likely to be overlap between packages prepared for vocational use in schools, with similar materials for use in industrial training or aimed at the home market.

17.6 An important area in the schools sector is special education (nearly 2000 schools), where although the problems that have to be faced are often individual and specific, the use of interactive video techniques may offer cost-effective solutions. This area should not be neglected in seeking applications, neither should an associated one of hospitalized children. The facilities for teaching children in hospital for long-term treatment are seriously limited and would be appreciably enhanced through imaginative use of interactive video.

Further and higher education

17.7 The arguments that apply to the potential of interactive video for schools have even more force in the further and higher education sectors. The emphasis on in-depth understanding of specialist subject matter and the encouragement of an individual rather than a group response is matched by the characteristics of the new medium. Programmes pertaining to almost all curriculum areas may be foreseen, ranging from introductory sequences, through specific course elements, applications exercises and review materials. The greater ability of students at this level to organize their own work with the help of tuition and counselling makes the possible availability of interactive video very attractive as a supplement to conventional texts and self-learning materials. The establishment of the Open Tech programme provides a further vehicle for the exploration of interactive video, and it is known that teams at the Open University are already attracted by the technology.

APPLICATIONS IN EDUCATION AND TRAINING

17.8 The developing pattern of further and higher education is of students coming from a wider and wider range of educational backgrounds and experiences, many returning to education after a break for work experience, or to reorient their careers. There is thus a growing need to gear course design and organization to reflect the characteristics and expectations of such students, who may not be best suited by the conventional linear, average-paced exposition of subject matter. Bork (1977) envisages an experimental course, making much use of interactive video, that would begin to meet such demands.

17.9 A series of multi-media learning modules would cover the initial stages of the course, with material at a variety of levels making a number of assumptions about, say, the mathematical background of the students. This multi-track approach will allow individual choice of content assisted by internal prescriptive decisions. Embedded testing procedures would indicate which objectives students were having difficulty with in order to offer immediate help. Responses to other tests would build up the student's credit rating.

17.10 Some discs would contain standard texts, cross-referenced, and updated with auxiliary material, giving in effect access to an electronic library. Others would have reference material, particularly graphs, illustrations and diagrams, which could have the added impact of full colour representation. Further discs would be concerned with background materials, putting the main subject matter into historical or contemporary context; others would contain enrichment materials allowing the individual student to explore avenues not in the normal direct line of the course objectives. More structured programmes would provide simulation or vicarious experiences not readily accessible in the laboratory or workshop through which mastery and understanding of procedural skills could be demonstrated. Additionally, the full facilities of a computer system would be available for student use, to amplify aspects of course work and for use in specific problem-solving.

INTERACTIVE VIDEO

17.11 Such a scenario may seem far-fetched *in toto*, but such elements judiciously integrated with the best of existing teaching practices could do much to liberate higher education from its conventional pedagogic straightjacket.

17.12 Interesting examples of the use of intelligent videodiscs are already beginning to appear in the field of post-experience medical education. These, aimed at the continuation training of GPs, or to disseminate information about new surgical techniques, could well be parallelled in a large number of professional fields.

Industrial and corporate training

17.13 Considerable expense and effort is devoted to vocational training within industry and commerce, either organized within individual companies, or provided on an industry-wide basis. Training schemes are also run within Government training establishments, and many further education establishments organize specific courses for local industry. The Manpower Services Commission now plays a very significant part in determining the extent and direction of industrial training initiatives, and its interest in innovative techniques has extended to support of computer-aided instructional techniques, for example use of the PLATO system for youth training in Coventry.

17.14 The potential of interactive video techniques in the vocational training field is immense, and is already being explored by leading companies. Possibilities of its use include the following.

17.15 *Operator training*. Basic manual skills, machine-operating practices, production processes, plant control operations, safety practices, etc, are all better learnt by training that 'shows how' rather than 'tells what'. Often trainees do not enjoy higher levels of literacy and the ability to absorb crucial information from the printed word alone. The ability of the videodisc to offer realistic pictures of situations, combined with its ability to respond patiently and endlessly to student inquiries, gives it a high rating as a training

APPLICATIONS IN EDUCATION AND TRAINING

tool in most fields where operator proficiency in standard procedures and practices is the prime objective.

17.16 *Supervisor training.* Much supervisor training involves learning rules and policies and how to apply them, whether these are in workshop or office practice, dealing with customers or with employees, or responding to management requirements. These contain a large element of procedural skills, juxtaposed with interpersonal skills. Supervisory training is often similar across one particular industry, and therefore common training elements can be devised. Again, the use of intelligent videodisc systems appears to offer significant advantages over conventional methods. It is not always convenient for staff to be released for block training and the freedom afforded by an individual learning system can be an added attraction.

17.17 *Sales training.* Basic selling techniques and customer handling can be simulated through videodisc presentations. Videodiscs are also invaluable for updating product knowledge, with opportunities for interactive indexing allied to pictorial representation of catalogue data. This parallels their use in direct sales communication and customer information services.

17.18 *Training of maintenance and repair staff.* Diagnostic and fault-finding training is highly procedural, and is often found at fault when the trainee returns to the field. Videodisc systems can not only assist in training sessions by simulating fault situations, but can offer assistance in the field in diagnosing symptoms and effecting cures.

17.19 *Management training.* Concerned largely with decision-making and development of interpersonal skills, a major problem in management training is lack of realistic practice. Interactive video offers the potential for problem-solving simulations and role-plays in which real people appear to participate. Particular attention could be given to the need to appraise management of the implications of information technology developments as they affect their particular industries.

17.20 In general, interactive video offers industry an opportunity to decentralize its training activities to branch or regional level, thus reducing costs and lost time, while retaining overall control of training practices and standards. This is particularly attractive to those industries with remote or travelling personnel — shipping, oil industries, etc.

Military training

17.21 Military training needs are similar in many cases to those of industrial training. There are equivalent areas of operator and maintenance training concerned with weapons and equipment, with the armed services frequently finding a high proportion of trainees with low reading abilities. At higher levels considerable use is already made of simulator training (eg, in pilot instruction), and the US military, in particular, is evaluating the potential of interactive video as a training system both for its own uses and as a means of training foreign purchasers of its equipment. An interesting possible application for intelligent videodiscs would be in the field of wargaming.

Adult/recreational education

17.22 Continuing education for adults is becoming an insistent theme. Already some 5500 adult education centres operate a range of part-time recreational and vocational courses for adults, but these fail to satisfy the demand, which may be expected to increase as the community comes to terms with changing work patterns. Supplementing the local authority provision is a growing commercial market for 'home' education, covering subjects such as cookery, carpentry and home maintenance, interior decoration, needlework, photography, recreational arts, motor mechanics, sports, and musical performance. Materials are also becoming available to assist the small businessman and to deal with personal finance. A further range of materials is aimed at areas of personal development, offering typically foreign language learning, and improvement of social and communication skills. This home market already embraces multi-media packages, and is rapidly involving the use of home computers.

APPLICATIONS IN EDUCATION AND TRAINING

17.23 The potential of videodisc technology in this market is already attracting the attention of the publishing and communications industries, although whether this will develop through linking home computers to consumer videodisc players, or through providing players with their own 'plug-in' intelligence (via ROM cartridges) on the pattern of many video games, is not yet clear.

18. POTENTIAL FOR COLLABORATION

18.1 A number of groups within the UK are already experimenting with aspects of interactive video, or are actively seeking information and resources to enable investigative work to begin. This interest has been stimulated by the recent launch of PAL videodisc equipment, and by the selective approaches of manufacturers, such as Philips, to assist in their product development.

18.2 Many of these groups have expressed interest in collaborative studies, and could be considered as potential contributors to any coordinated research and development programme. The following list, which includes some that have already been mentioned, is by no means exhaustive, but illustrates the wide range of concern in the educational and training fields.

Further and higher education

18.3 The Department of Maritime Studies, University of Wales Institute of Science and Technology, has been developing computer-based training packages for ocean-going merchant fleets, in cooperation with Texaco Overseas Tankship Ltd, and is now extending this activity to introduce videodisc technology. Associated applications of videodisc relate to provision of navigational charts and aids, and engineering drawings and documentation, necessary for the safe operation and handling of merchant vessels. Another group at UWIST is concerned to employ videodisc techniques in the teaching of English as a foreign language.

18.4 The Department of Zoology at University College, Cardiff, is concerned to update a self-learning course in the first year, and incorporate appropriate video material.

18.5 At Dundee College of Technology, work is in hand to employ high-resolution colour pictures in a computer-aided learning scheme in biochemistry. Experiments with random-access slide equipment, and with videotape machines point to the greater effectiveness expected with videodisc. This work has been supported in part by the Scottish Education Department.

18.6 The London University Audio-Visual Centre is bringing together interactive video interests within the University and programmes in the fields of medicine and law are expected to be undertaken. Further work is in hand to convert Meteosat pictures of Earth to videodisc format in cooperation with the National Aeronautics and Space Administration.

18.7 A group in the Institute of Educational Technology at the Open University is studying the pedagogic implications of interactive video for open learning, first using videotape technology and now videodisc. Initial work has been done in connexion with a signal communications module of a basic technology course. A visiting American expert from the University of Nebraska is currently working with the group. The BBC/OU Production Unit also has interests in videodisc technology (see below).

18.8 Work in the Design Research Department of the Royal College of Art has concentrated on the development of database structures for the design professions, through the integration of computer and videodisc technologies. This group has links with the Architecture Machine Group at Massachusetts Institute of Technology.

18.9 At Hatfield Polytechnic, the Department of Engineering has been studying the potential of optical disc technology for information storage and retrieval, with support from the British Library Research and Development Department. A project to develop a videodisc programme is in hand.

18.10 Associated work in the use of microcomputing in libraries, with concern for visual information and user education, is being undertaken at the Polytechnic of Central London, again under the auspices of the British Library.

INTERACTIVE VIDEO

18.11 Brighton Polytechnic is planning to undertake a project to link videotex and videodisc technologies, and seeking applications in the travel agency business and in legal information searches. Although not directly educational, the basic technology involved will be similar and a further application is in mind in connexion with the South Tech training scheme proposed to the Manpower Services Commission as part of the Open Tech programme.

Public corporations

18.12 Milton Keynes Corporation commissioned the BBC/Open University Production Unit to produce a videodisc to demonstrate how access to multichannel cable networks and satellite broadcasting would affect the provision of video, and how interactive services could be provided. The opportunity was taken to illustrate the potential the medium offered for both recreational and formal education. The Production Unit is perhaps one of the best equipped to produce interactive video materials.

18.13 BBC Enterprises was responsible for the first British commercial interactive videodisc, 'British Garden Birds'. The Corporation is currently assessing its future role in the production and distribution of interactive video programmes.

18.14 It is understood that British Telecom is investigating the application of interactive video in various training situations.

Industrial training

18.15 Interests in industrial training and commercial applications of interactive video can be divided between consultant companies and facilities houses capable of designing and producing materials for specific customers, and those companies who support their own in-house training facilities.

18.16 Examples of the former, which are specifically promoting their expertise in interactive video, include EPIC, FELIX, Scicon-CAVIS, Mitchell Beazley, Butler Cox Associates,

POTENTIAL FOR COLLABORATION

Realmheath, Video Arts Molinaire, TVI, etc. To these should be added Thorn-EMI which, although it has postponed the launch of its own videodisc system, has retained its interactive video programme development team. Some of these companies have partnerships with US agencies, or are able to tap other production expertise.

18.17 Smith, Klein and French Laboratories have commissioned EPIC to anglicize a number of programmes produced in America aimed at updating general practitioners in modern medical techniques. These will be used by SKF in their own educational promotions. Longman have associated with Thorn-EMI in developing a trial disc on teaching to play the recorder. Such a programme could be aimed at schools or for the home market. Another recreational consumer disc has been prepared by JDF Associates with Thorn-EMI on instructing fishing techniques. It is understood that Macmillan are also interested in publishing videodiscs. Wiley Educational Software is publishing an American disc, 'The Puzzle of the Tacoma Narrows Bridge Collapse' and plans to follow this with packages designed to meet British educational requirements.

18.18 In-house investigation of interactive video is being undertaken by a number of companies in widely varying industries. British Leyland Systems are investigating the application of videodisc technology, possibly in conjunction with viewdata technology, in areas such as archival storage of drawings, marketing and dealer education, spare parts and stock control, and are considering offering training and instructional design services based on interactive technology. Similar interests have been expressed by Ford (GB) Ltd, particularly in the field of workforce training, where new production processes require higher levels of understanding and skill than hitherto in setting and interpreting instruments. Marconi is believed to be evaluating advanced simulation systems using videodiscs for technical training. Also in the engineering field, the Engineering Industry Training Board has long provided training manuals and 'teach yourself' materials on an industry-wide basis in conjunction with their module system of training. This subject matter frequently contains material that demands illustration, and some interest has been expressed within EITB in the potential of interactive video to extend the effectiveness of such programmed guides.

18.19 The clearing banks already employ video techniques on a large scale in their training and information dissemination programmes. Barclay's have over 1400 VTRs at branch level, and National Westminster over 1000. Barclay's are evaluating interactive video. In a related area the Chartered Building Society Institute is concerned to evaluate the value of interactive video as an industry-wide training technique. Commercial Union Assurance Co has also expressed interest.

18.20 In the retail field videodisc technology is already being employed in point-of-sale customer education (Mothercare), and ABTA are backing an investigation of the value of linking videodisc to private viewdata booking systems for travel agents. Other retailing multiples with extensive in-house training programmes are considering the use of interactive video (Marks and Spencer, John Lewis, etc).

Military training

18.21 All three armed services have groups interested in interactive video developments as they may affect their own training requirements — the Royal Navy School of Educational and Training Technology, Portsmouth; the Army School of Training Support, Beaconsfield; and the RAF Support Command HQ at Brampton. Specific use of some techniques is already being made, eg, use of the Sony Responder by the Royal Military Police.

19. EXECUTIVE SUMMARY AND RECOMMENDATIONS

19.1 There has been continuous progress in the application of television to education and training. New equipment and facilities now allow television to move beyond its use as a mass communications medium to serve the needs of individual learners. A stage has been reached where the possibility of a learner, assisted by microcomputer technology, 'interacting' with video material is demonstrable (see 1.5).

19.2 Interactive video offers unique opportunities to exploit electronic technology to support a range of educational and instructional tasks (see 2.2). Moreover, the commercial prospects for a new 'knowledge-based' industry are worth exploring (see 2.5). This study attempts to identify the features and advantages of the system, and to propose a course of action to promote its development.

19.3 Broadcast television alone cannot offer individual control over when, where and how the medium is used. Programmes are transmitted at fixed schedules; are largely holistic in nature; and aim at the average viewer as perceived by the producer (see 3.2). These characteristics are largely common to other distribution services, such as cable systems. Although the advent of recording equipment offers a greater degree of control by the user, the extent of 'random-access' offered is still limited (see 3.4).

19.4 Computer-based learning systems can offer a range of individual educational experiences, but have tended to provide only clumsy visual displays that inhibit full acceptance of the technology. Resolution of the issue of combining computer generated images with high-quality television pictures is now within reach (see 3.8).

INTERACTIVE VIDEO

19.5 The application of other information technologies to education is just beginning to become appreciated. Contact between teacher and taught and student access to databases via telecommunications links offer significant advantages, as do use of facsimile, videotex and teletext technologies (see 3.11). These systems can all be linked to enhance the value of interactive video.

19.6 The unique character of interactive video is that it can offer all the attributes required of an educational communication system, it can encompass a variety of different teaching and learning strategies, it can monitor student performance and adapt to individual needs, and it can provide the essential motivation (see 4.5).

19.7 Equipment is becoming available which links a videocassette recorder to a microcomputer system, either as a custom-built unit or as component parts. These offer access to videotaped sequences to extend standard computer-aided learning practices, but most rely on elementary multiple-choice decision-making. VCRs have considerable limitations when extensive searching is required or when single images are appropriate. Although valuable to encourage experimentation and supportive of local programming, the opportunities for exchanging courseware between systems is slight (see 5.7). A prototype modular system may cost £3000 (see 5.17).

19.8 Advances in laser technology have made possible optical recording systems that offer very high density information storage at reasonable cost (see 6.2). The videodisc is being developed in numerous forms, but the only current operational system suitable for educational interactive video is the LaserVision system supported by Philips, Pioneer, Sony, and associated companies (see 6.15).

19.9 The LaserVision system offers some half-hour of playing time (54,000 frames) on each side of a rugged, read-only, 30cm diameter record (see 7.4). The signal is read from the disc by a low-power laser, and it can be operated in forward

EXECUTIVE SUMMARY AND RECOMMENDATIONS

and reverse, fast and slow, and in stop-frame modes as well as normally. Players can quickly jump from one picture sequence to another, either in response to pre-ordained commands or under control of an external computer. Players of varying degrees of sophistication are becoming available, offering different levels of 'intelligence' and function (see 7.9). Equipment to the PAL specification has only recently become available.

19.10 Domestic equipment for the consumer entertainment market can be adapted to respond to computer signals and used for significant interaction (see 8.3). However, it is preferable to have equipment designed to 'industrial' standards. Interface equipment is necessary to link videodisc players to common microcomputers, and to control other functions, such as switching monitor output between computer and player (see 8.10).

19.11 The use of standard videotex generators is a convenient way to incorporate text with video images. But the limitations on quality and resolution of videotex images suggests that care should be taken not to condition the development of interactive systems by too heavy reliance on current videotex technology (see 9.7). Linking interactive video systems with gateway viewdata services could be particularly fruitful (see 9.11).

19.12 It is already apparent that different equipment suppliers are taking different approaches to the design of control and command protocols, and that these divergences will create difficulties for the universal interchange of programmes. It would be helpful if some common specifications could be agreed, at least among educational and instructional users (see 10.8). The creation of teaching programmes could also benefit from the development of powerful authoring systems, preferably where the logical and content elements are separately defined (see 10.9). Additional software development is required to afford students access to a full range of computer-based facilities (see 10.12).

INTERACTIVE VIDEO

19.13 Although existing television material may be readily transferred to videodisc, it is necessary to ensure that the necessary copyright clearances can be forthcoming, and it is hoped that owners of material will not be too restrictive (see 11.1). However, existing material should be used selectively since the new medium offers the opportunity of a completely different design philosophy (see 11.2).

19.14 The enormous capacity of the disc to store still images suggests a possible use as a resource bank for visual data. This will require considerable work on problems of indexing, and will need to operate within the resolution limits set by broadcast television technology (see 11.7). Storage of text and documents also suffers from the same problems. Careful consideration should be given to the relative advantages of using digital means (computer and magnetic store) or analogue means (signals on videodisc) to hold text, computer programs, audio signals, etc, and to see the system as a closely integrated and optimized information and communication device (see 11.13).

19.15 Although it is inevitable that in the early stages of development emphasis will be placed on technical issues, the successful acceptance of interactive video systems will be largely influenced by the range and quality of the programmes — courseware — made available. It is crucial to build its applications around the best practices that have already evolved within the constituent educational technologies, although these in themselves will need considerable adaptation (see 12.2).

19.16 The videodisc can be employed as a universal audio-visual medium, offering any combination of still and moving pictures with or without sound. Combined with a computer system, a user may be offered full access to computational facilities and to both local and distant databases (see 12.7). Given this versatility, interactive video can equally serve a variety of instructional strategies (see 12.9).

19.17 Not enough is known yet about the most effective means of achieving educational ends, and the application of any

EXECUTIVE SUMMARY AND RECOMMENDATIONS

educational tool has to evolve pragmatically. Interactive video could perhaps be most significant in advancing the teaching of procedural skills, as against mere declarative knowledge, to exploit the ability of the human brain to process complex visual images, and to add interpersonal and emotional elements to assist attention and focus learning (see 12.14). Interactive video can accommodate a range of different learning styles and may lead to greater understanding of the psychological processes of learning (see 12.17).

19.18 It is clear that the technical features of the videodisc process require professional standards of programme production (see 13.3). Original material can be assembled from film, tape or slide, but it is likely that satisfactory results will only be obtained through employment of computer-controlled editing techniques, and electronic image handling (see 13.8). Production is therefore a relatively expensive process.

19.19 The design and production of an interactive video programme demands the integration of a number of professional skills, and the most successful efforts are likely to be team enterprises (see 13.11). Authoring can be expected to flourish best in an environment with considerable logistic support and where concentrated effort can be brought to bear on the task. It is possible, however, to consider the development of basic interactive 'resource' compilations, which will allow local teachers to create specific teaching programmes with the aid of simple do-it-yourself authoring systems (see 13.13).

19.20 Formative evaluation of programme material during the production stage is recommended, since changes cannot be accomplished once mastering is begun. This can be assisted by access to a simulation facility based on a random-access editing videotape recorder (see 13.15). The costs of mastering discs are of the order of £2000, with a charge of about £10 for each replicate produced. However, when all costs of design and production are taken into account, the cost per programme could average £50,000 (see 13.22).

INTERACTIVE VIDEO

19.21 The resource requirements for design and production of interactive video materials are extensive. Manpower contributions are necessary from the fields of subject expertise, teaching and curriculum development; instructional design and evaluation; text editing and graphic design; computer programming and systems engineering; film and television production; and project direction and management (see 14.2). For most economic operation development teams should be associated with relatively few production centres capable of providing sufficient logistic support (see 14.4).

19.22 A major requirement is access to broadcast standard television production facilities. However, there are a number of existing studio facilities in the education service that could be suitably upgraded (see 14.7). This is considered preferable to the *ad hoc* use of facilities houses, although it is recommended that management of studio facilities should be exercised on commercial lines (see 14.9).

19.23 Comprehensive computing facilities also need to be provided for production teams (see 14.11). These should be compatible with microcomputers commonly available in schools and colleges, but should also include examples of 'new technology' micros, particularly those capable of supporting an extensive authoring system (see 14.13). Separate computer systems should be provided for systems and software development, and modular, machine-independent, principles should be adopted in software design (see 14.14).

19.24 Careful consideration should be given to the most appropriate design of user terminals, bearing in mind that very little PAL compatible equipment is currently available (see 14.15). A target cost of less than £2000 is achievable.

19.25 The scale and range of resources necessary to make any real progress is such that special funding arrangements will need to be made for the initial research and development stages (see 15.2). It is likely that many different government departments and agencies will be pressed to give support to

EXECUTIVE SUMMARY AND RECOMMENDATIONS

investigative projects, and indeed some are already doing so (see 15.3). There is danger that a piecemeal approach will be of limited effectiveness and could lead to wasteful duplication of effort. A coordinated programme of research and development is thus recommended (see 15.5).

19.26 It is proposed, however, that only part of the funding for such a programme should come from government and that significant industrial and commercial involvement is sought from the outset. The reasons for this are that the techniques can equally benefit education and industrial and corporate training needs; that developmental collaboration between sectors of industry and the formal education system is to be welcomed; that the programme should be definitely applications oriented; and that an objective should be the creation of a viable 'knowledge-based' industry (see 15.6).

19.27 A consortium approach is therefore suggested with the following components: government interests; electronics industry; computer industry; publishing and distribution industry; television and communications industry; merchant banking interests (see 15.7). A scale of activity should be established aimed at reaching a position by the end of the third year whereby decisions on the future viability of the 'industry' could be made (see 15.8).

19.28 A model research and development programme is suggested, with the following objectives: to develop the technical base and establish necessary facilities; to create a series of teaching and instructional programmes; to establish the basis of a new 'knowledge industry'; to provide training and information for users; and to establish specifications and standards (see 16.1). Five major centres should be identified to undertake production work and provide services for authors and course designers. The programme should run three years in the first instance, by which time a target of some 80 productions should have been completed, a proportion of which would be revenue-earning (see 16.6).

INTERACTIVE VIDEO

19.29 Until it is possible to launch an integrated research and development programme, an interim user forum should be established, possibly under the auspices of CET, to deal with urgent user issues and maintain a watching brief on developments (see 16.8).

19.30 A wide range of applications for interactive video in all sectors of education and training can be identified. In primary schools the main emphasis is likely to be on basic skills (see 17.3). In the secondary sector applications are forseen in many subject areas, particularly as resource compilations (see 17.4), and in the sciences and vocational training (see 17.5). There are likely to be further applications in respect of special education, and in the case of hospitalized children.

19.31 The further and higher education sectors are perhaps the first areas of education to accept interactive video, giving opportunities for students to organize their own work. In the field of open learning both the Open University and the Open Tech can be expected to show great interest in interactive video (see 17.7). Applications in post-experience education and continuation training, for example in the medical field, are also likely to prove fruitful (see 17.11).

19.32 In industrial and corporate training, possibilities exist in areas such as operator training, training of supervisors and sales staff, training of maintenance and repair personnel, and in management training (see 17.14). The technique could prove particularly attractive to companies with dispersed operations, eg, shipping (see 17.20).

19.33 Many military training applications can also be forseen, including field training, and training foreign purchasers of equipment (see 17.21).

19.34 Interactive video is felt to be an attractive medium in the growing commercial market of 'home' education. Applications are expected in many areas of recreational and

EXECUTIVE SUMMARY AND RECOMMENDATIONS

vocational skills, and also in areas of personal development. The potential link to home computers is particularly interesting (see 17.22).

19.35 A number of groups in the UK are already experimenting with interactive video. These include teams in universities and polytechnics, companies concerned with industrial training, public corporations, and publishers, as well as those companies directly involved in developing the hardware. Many of these groups have expressed interest in collaborative studies, and could be considered potential contributors to any coordinated research and development programme (see 18.2).

REFERENCES

BACKER, D, 'One-of-a-kind video programs', *Instructional Innovator*, 26, Feb 1982

BARRETT. R, 'Prospects for the optical disc in the office of the future', *Reprographics Quarterly*, 14 (4), 1981

BARRETT, R, *Developments in Optical Disc Technology*, British Library Research and Development Report 5623, 1981

BARRETT, R, *Optical Videodisc Technology and Applications*, Library and Information Research Report 7, British Library, 1982

BORK, A, 'The educational possibilities of intelligent videodiscs', Proceedings of the Annual Conference of the Association for Computing Machinery, p271, 1977

BRYCE, C, *Improved CAI by the Use of Interfaced Random-Access Audio-Visual Equipment*, Dundee College of Technology Research Report P/24/1, 1982

CIARCIA, S, 'Build an interactive videodisc controller', *Byte*, 7 (6), p60, 1982

COPELAND, P, 'Study by video: learning made easier', *Journal of Educational Television*, 7 (1), p7, 1981

DAYNES, R, 'Experimenting with videodisc', *Instructional Innovator*, p24, Feb 1982

DAYNES, R, 'Videodisc interfacing primer', *Byte*, 7 (6), p48, 1982

HON, D, 'Interactive training in pulmonary resuscitation', *Byte*, 7 (6), p108, 1982

HUFSCHMID, P et al, 'Experimental teaching program for interactive microcomputer-managed videodisc player', Council of Europe Conference on New Communication Techniques in Post-Secondary Education, Strasbourg, Sept 1979

KENNEY, G C, 'Special-purpose, applications of the Philips-MCA Videodisc System, *IEEE Transactions on Consumer Electronics*, p104, Nov 1976

REFERENCES

KEWNEY, G, 'Interaction in perspective', *Corporate Video,* Oct 1981

LAURILLARD, D M, 'The potential of interactive video', *Journal of Educational Television,* 8 (3) p73, 1982

O'NEAL, F and LIPSON, J, 'Instruction with the intelligent videodisc', ASEE Annual Conference Proceedings, 1982

REEVEL, P, 'Making videodiscs today', *Audio Visual,* p36, July 1982

SCREEN DIGEST *'Videodisc status report', Screen Digest,* p109, June 1982

WOOD, R K with WOOLLEY, R D, *Videodisc Technology: applications to library, information, and instruction sciences,* ERIC Clearinghouse for Information Resources, Syracuse University, New York, December 1980

APPENDIX 1. DIGITAL DATA DISCS

A1.1 Barrett (1981, 1982) has produced comprehensive surveys of optical disc technology, with an emphasis on the applications to high resolution image/document storage.

A1.2 Kenney (1976) had earlier described a number of ways in which the LaserVision videodisc could be used to store special-purpose information in applications such as digital read-only memories, X-ray and document storage, and as a talking encyclopedia.

A1.3 The television signal format, as used on the optical videodisc, can be adapted for data applications as shown in Figure A1.1. The standard NTSC signal has picture information interrupted every line by synchronization signals. (PAL signals are similar.) Useful information is present for only part of each line, and it is into this time-slot that data signals can be inserted. Data signals must therefore be divided into discrete packets by suitable buffering circuitry at recording, and equally decoded on playback.

Digital read-only memory

A1.4 Since the NTSC signal carries a colour burst at 3.58 MHz it is convenient to choose record data at a rate of 7.16 Mbits/sec. This yields a capacity of 375 bits per line, or 185,625 bits per frame. Utilizing all 54,000 frames yields a total storage capacity of some 100 Gigabits per videodisc side, or 1250 million bytes.

A1.5 Each track on the disc will have a separate address, and thus reading information is similar to that for a magnetic disc. The player tracking mirror can address a band of 100 tracks within a few microseconds. Such a band has a capacity of over two million bytes, roughly equivalent to that of a typical magnetic disc-drive, but at far lower cost. The

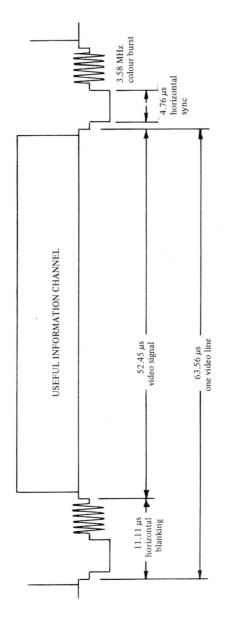

Figure A1.1. NTSC colour video signal

INTERACTIVE VIDEO

optical videodisc has therefore significant potential as a read-only memory in conjunction with a digital computer system.

High-resolution image store

A1.6 Where it is necessary to store images at a resolution greater than the capability of the standard television signals either the image can be subdivided, or recourse made to a slow-scan method. A normal NTSC frame can display 420 horizontal picture elements (pixels) and 475 vertical pixels, or 0.2 million pixels per frame. A typewritten A4 page contains some 1.6 million pixels and will thus require 8 full television frames to be satisfactorily reproduced. A 2000-line resolution photograph will require 40 million pixels, or 20 television frames. Simple multiframe recording techniques divide large images into a matrix of sub-pictures, but these often lead to problems during reproduction due to overlap of detail at the matrix boundaries.

A1.7 By using a slow-scan technique it is possible to arrange that each line contains the necessary number of pixels. Thus, if a 2000-line image is scanned at one-fifth the rate of the normal TV horizontal scan, each picture line can be converted into five TV lines. Continuing this way it is possible to contain 2000 vertical lines within 19 normal NTSC frames. These can be conventionally recorded and retrieved, and the original high-resolution image recovered using a scan conversion storage device.

Optical digital data discs

A1.8 An important factor affecting the use of standard videodiscs for data recording is the rate of occurrence of errors introduced by the disc mastering and replication process. Random 'drop-outs' produced by the presence of dust and dirt particles have often little effect in television recordings due to considerable redundancy in successive frames and persistence of vision. However, obliteration of blocks of digital information can result in appreciable errors, and the need to overcome these problems has hindered the digital application of videodiscs.

APPENDIX 1

A1.9 In an attempt to meet the requirement for low error rate a number of companies are investigating direct-read-after-write (DRAW) laser recording systems, which allow the recorded data to be checked instantly so that any faulty data can be immediately re-recorded. The Philips DRAW system is based on a tellurium-coated disc, on which 40,000 tracks can be recorded giving a storage capacity of 100 Gigabits per side. RCA have been experimenting with titanium sub-strates; Bell Laboratories with bismuth. The Japanese Toshiba company offer a fast-access electronic document store based on work undertaken at Stanford Research Institute. Using a tellurium-coated disc, 2000 pages can be recorded on each side at 215 points/in resolution. The archival qualities of such media are not yet totally satisfactory, and it is expected to be some years before digital optical disc systems seriously challenge magnetic recording media.

APPENDIX 2. LASERVISION SYSTEMS: PROVISIONAL DESCRIPTION OF COMPUTER INTERFACE FOR USE WITH THE PHILIPS PROTOTYPE INDUSTRIAL PLAYER

The Council thanks Philips Electronics for their permission to use this material.

1. General

The player is equipped with an external control bus. The control bus will operate in accordance with the RS 232-C standard.

The LVS-player will have a 25-pole female connector with the following pinning.

Pin 7   Signal ground
Pin 2   Transmitted data from LVS player to external computer
Pin 3   Received data from external computer to LVS-player
Pin 5   Clear to send
Pin 20 Data terminal ready

Transmission speeds are 1200 or 9600 baud. The speed is selected by means of a slideswitch located on the print panel under the top left ornamental cover.

SWI ON = 1200 baud. SW1 OFF = 9600 baud.
Note: SWI is factory set to 1200 baud.

The higher baud rate gives an improvement if teletext commands are used. Teletext with 9600 baud will be about five times faster than 1200 baud.

Data format is one startbit, eight databits plus one stopbit. No parity bit.

With the second slideswitch on the print panel the 'handshake' information on pin 5 can be switched off.

Note: SW2 OFF = no handshake SW2 ON = handshake given.

APPENDIX 2

A message to the LVS-player should consist of a single ASCII-code plus carriage return "CR" or a string of characters plus "CR". No line feed. Action in player will start after "CR". The maximum length of a string including "CR" is 64 characters. A longer string will be rejected.

ASCII-codes lower than 32 and higher than 127 will also be rejected. If a string has one character lower than 32 or higher than 127 except "CR" the whole string will be rejected.

From the LVS-player 06 will be sent back as an acknowledge on all strings beginning with T or P. For other strings no acknowledge will be sent out. (See command list.)

2. Command table

	HEX	DEC	CHAR	SHORT FORM	TIME
2.0	2E	46	.	CORRECTION	
2.1	2F	47	/	ENTER	
2.2	30	48	0	DIGIT 0	
2.3	31	49	1	DIGIT 1	
2.4	32	50	2	DIGIT 2	
2.5	33	51	3	DIGIT 3	
2.6	34	52	4	DIGIT 4	
2.7	35	53	5	DIGIT 5	
2.8	36	54	6	DIGIT 6	
2.9	37	55	7	DIGIT 7	
2.10	38	56	8	DIGIT 8	
2.11	39	57	9	DIGIT 9	
2.12	3A	58	:	PAUSE	
2.13	3B	59	;	MEMORY	
2.14	3C	60	<	SEARCH REVERSE	
2.15	3D	61	=	AUTO-STOP	
2.16	3E	62	>	SEARCH FORWARD	
2.17	3F	63	?	PICTURE NUMBER	
2.18	40	64	@		
2.19	41	65	A	AUDIO I: A1=ON A0=OFF	200 ms
2.20	42	66	B	AUDIO II: B1=ON B0=OFF	200 ms

INTERACTIVE VIDEO

2.21	43	67	C	CHAPTER NUMBER: C1=ON C0=OFF	200 ms
2.22	44	68	D	PICTURE NUMBER: D1=ON D0=OFF	200 ms
2.23	45	69	E	VIDEO MUTE: E1=ON E0=OFF	40 ms
2.24	46	70	F		
2.25	47	71	G		
2.26	48	72	H	REMOTE EXT: H1=ON H0=OFF	40 ms
2.27	49	73	I	KEYBOARD ACTIVE: I1=ON I0=OFF	40 ms
2 28	4A	74	J	REMOTE TO INT.: J1=ON J0=OFF	40 ms
2.29	4B	75	K	KEY RELEASE	40 ms
2.30	4C	76	L	STILL FORWARD	100 ms
2.31	4D	77	M	STILL REVERSE	100 ms
2.32	4E	78	N	NORMAL PLAY FORWARD	200 ms
2.33	4F	79	O	NORMAL PLAY REVERSE	200 ms
2.34	50	80	P	GO-TO PICTURE NUMBER	-
2.35	51	81	Q		
2.36	52	82	R	RUN	40 ms
2.37	53	83	S	SLOW SPEED CHANGE	40 ms
2.38	54	84	T	TELETEXT	60 ms
2.39	55	85	U	SLOW MOTION FORWARD	200 ms
2.40	56	86	V	SLOW MOTION REVERSE	200 ms
2.41	57	87	W	FAST	40 ms
2.42	58	88	X	CLEAR	40 ms
2.43	59	89	Y	REPEAT	40 ms
2.44	5A	90	Z	PLUS	40 ms

Note: the timetable expresses the time consumption for the player to execute different commands. If no handshake is given, delays are needed in the computer program.

3. Comments to command table

3.1 One byte commands with no acknowledge back

3.1.1 Search reverse, search forward and fast. These three commands are special as they start either a search or fast

APPENDIX 2

mode in the LVS-player and to stop that function "K" = Keyrelease must be sent. This can be compared with the operation of the keyboard on the player where search and fast are active as long as the button is pressed.

3.1.2 Still forward and still reverse. On receipt of these commands the player will go into the still mode. If the player is in the still mode the player will step one picture forward or reverse one picture.

3.1.3 Normal play forward and reverse. These commands will start normal play forward or normal play reverse.

3.1.4. Slow motion forward and reverse. These commands will start slow motion either forward or reverse. For speed selection see 3.3.

3.1.5 Enter, correction, digits, memory, auto-stop, run, clear, repeat and plus. All these commands are normally sent by the remote control transmitter and should not be normally used as commands sent by the external computer.

3.1.6 Pause. This command has a toggle function and is normally used only from the remote control transmitter. Status ON or OFF cannot be seen from the external computer.

3.1.7 Picture number. After receiving this command, the player will send back the actual picture number. If no handshake is used the answer will come back after 300 ms or faster.

3.2 Two byte commands with no acknowledge
3.2.1 Audio I ON or OFF and Audio II ON or OFF. These commands control the audio channels. If a command is sent to the player while the player is in, eg, still mode the command will be stored and executed when the player returns to the normal play forward mode.

135

INTERACTIVE VIDEO

Note: after starting the player an "AX" and a "BX" (X = I or O) should be sent to obtain the wanted channels and correct setting of the memories in the LVS-player.

3.2.2 Chapter display ON or OFF. This function is present if there are chapter numbers on the disc.

3.2.3 Picture number display ON or OFF. This command will display or not the actual picture number.

3.2.4 Remote to external ON or OFF. If an "HI" is received the player will start the transfer of all remote transmitter commands received to the external computer. See 4 for a list of those commands. Note that an "HI" will generate a "JO" inside the player so that no remote commands will be executed by the player itself. This means that after sending an "'HO" a "JI" must be sent to restore control by the remote transmitter. Default after the start is "OFF".

3.2.5 Keyboard enabled or disabled. With these commands the local keyboard on the LVS-player can be enabled or disabled. No commands can be sent back to the computer. Default is ON.

3.2.6 Remote to internal ON and OFF. These commands enable or disable the remote transmitter commands to the LVS-player. See also 3.2.4.

3.2.7 Video mute ON or OFF. Suitable to use in still picture mode to give a blank screen. When used in the normal play mode the sound channels will also be muted.

3.3 Slow motion speed command "S" + "digits" + "CR"
The range is from S2 + "CR" up to S 255 + "CR".
S2 is normal speed — 25 pictures/sec
S50 is — 1 picture/sec

By multiplying the digit value by 20 ms the delay between

APPENDIX 2

each new picture can be obtained; eg, S50 gives a delay of 50 x 20 ms = 1 sec between each new picture. No acknowledge is given.

3.4   Commands with acknowledge

3.4.1 "P" + "digits" + "R" + "CR". Command for "looking up" a picture number. Digits can be from one digit up to 5 (eg, P5R, P325R, P23547R). On receipt of a command the player will mute video and sound and start searching for the wanted picture number. When the correct number is reached the picture will be shown and an acknowledge 06 will be sent back to the computer.

3.4.2 "P" + "digits" + "S" + "CR". Digit range same as 3.4.1. This command shall be used to stop the player on a required picture number from normal play or slow motion mode. When the required number is reached the player will go into the still picture mode and send back 06 to the computer.

If the player cannot find the wanted picture number but reads a picture number higher or lower than the wanted one it will send back 15H = "NAK". This will enable the computer to compare the picture number read with the wanted one and thereby tell the player to move forward or back by the required number of pictures and therefore display the wanted one.

3.4.3 "P" + "digits" + "I" + "CR". "I" stands for information and the player will send back an 06 if the wanted number is read.

The player will not change mode.

This command is suitable for subtitling with teletext. If something is wrong with that number 15H = "NAK" will be sent back. (See 3.4.2.)

INTERACTIVE VIDEO

4. List of codes for remote control transmitter command back to external computer if the LVS-player has received "HI"

Transmitter button		Code to external computer
Normal play	Forward	4EH = N
Normal play	Reverse	4FH = O
Still	Forward	4CH = L
Still	Reverse	4DH = M
Search	Forward	3EH = >
Search	Reverse	3CH = <
Slow motion	Forward	54H = T
Slow motion	Reverse	55H = U
Slow speed	plus	48H = H
Slow speed	minus	47H = G
Audio I		41H = A
Audio II		42H = B
Pause		56H = V
Fast		57H = W
Digit 1		31H = 1
Digit 2		32H = 2
Digit 3		33H = 3
Digit 4		34H = 4
Digit 5		35H = 5
Digit 6		36H = 6
Digit 7		37H = 7
Digit 8		38H = 8
Digit 9		39H = 9
Digit 0		30H = 0
Enter		45H = E
Correction		43H = C
Picture number		44H = D
Chapter number		51H = Q
Memory		50H = P
Run		4BH = K
Clear		58H = X
Auto-stop		53H = S
Plus		5AH = Z
Repeat		46H = F

Note: every code back to the external computer will be followed by "CR".

APPENDIX 2

5. Teletext commands

5.1 Page selection command
"T/PXXX" XXX are three digits in a range from 100-899 and following teletext headers and row-commands will be automatically sent to that page number.

Note that after a page selection command a header must be sent to restore the text-tv.

The command is used if different teletext, eg, two languages are displayed on two television sets with the same video background information from the player.

5.2 "Header commands"
5.2.1 "T/HBL". (T slash HBL) will erase the old teletext information and give a black picture without video from the player.

5.2.2 "T/HBN". Same as 5.2.1 but without erase of the old teletext content.

5.2.3 "T/HCL". Will give a "clear" picture with video visible. Old teletext information will be erased. Text must be inserted in subtitle mode ("start box" command must be used).

5.2.4 "T/HCN". Same as 5.2.3 but without erase of the old teletext content.

5.2 "ROW" commands
22 rows are used 1-22 and syntax is T/RXX:

    Examples.   T/R05 means row 5
                       T/R15 means row 15

Row command must be followed by "control" commands or text.

5.4 "Space" commands
Syntax is /XX. XX is a value between 01 and 39. The command is used if a space is wanted between two characters.

139

INTERACTIVE VIDEO

5.5   "Control" comands
5.5.1   "DH" = double height
5.5.2   "NH" = normal height
5.5.3   "FL" = flashing character
5.5.4   "ST" = steady
5.5.5   "SB" = start box
5.5.6   "EB" = end box
5.5.7   "CR" = colour red
5.5.8   "CG" = colour green
5.5.9   "CY" = colour yellow
5.5.10 "CB" = colour blue
5.5.11 "CM" = colour magenta
5.5.12 "CC" = colour cyan
5.5.13 "CW" = colour white
5.5.14 "GR" = graphics red
5.5.15 "GG" = graphics green
5.5.16 "GY" = graphics yellow
5.5.17 "GB" = graphics blue
5.5.18 "GM" = graphics magenta
5.5.19 "GC" = graphics cyan
5.5.20 "GW" = graphics white
5.5.21 "SG" = separated graphics
5.5.22 "HG" = hold graphics
5.5.23 "RG" = release graphics
5.5.24 "NG" = normal graphics = contiguous graphics
5.5.25 "BB" = black background
5.5.26 "NB" = new background
5.5.27 "CD" = conceal display

5.6   "Text" information
Text information visible on the screen shall be surrounded by the syntax "@".

5.7   Examples
5.7.1 "T/HBL" gives a "black" screen.
5.7.2 "T/R10/CR @ TEST @ gives the word TEST in red letters on row 10.
5.7.3 "T/R05/DH/15 @ ZERO @" gives the word ZERO in double height 15 spaces in on line 5 + 6. (Both lines due to command CH.)

5.7.4 "T/R20 @ START @ 10 @ END @ " gives the word START in the beginning of line 20 and the word END 10 spaces in on that same line.

5.7.5 "T/HCL" gives a "clear" picture, all following strings must use the command :"SB".

5.7.6 "T/R10/SB@INFORMATION@EB/" gives the word INFORMATION on a black background. If EB is not used the box will extend to the right side of the screen.

5.7.7 "T/R20/DH/CG/NB/CR/SB @ DIGIT @ FL @ 5 @ EB/" gives the word DIGIT in double height coloured red on a green background and a red 5 flashing.

5.7.8 Note that maximum length of a string is 64 bytes including carriage return.

APPENDIX 3. LASERVISION PROGRAMME MASTER TAPE SPECIFICATION (PAL 625/50) DECEMBER 1982

The Council thanks Philips Electronics for their permission to use this material.

1. TV standard
PAL 625 lines/50 fields.
(acc. CCIR report 624-1 systems PAL "B", "G" and "I".)

2. Tape system
2in transverse-track (Quad). acc. IEC 347. (High speed only.)
or
1in type "C" Helical. acc. IEC draft 60B-40. (With exception of control track level. See Item 4.)

3. Time errors
Max. 6nS pp.

4. Control track
Should start from the *first second* of tape lead-in, with *no* breaks, or discontinuities, right through to the last second of tape lead-out.
The 8-field PAL colour sequence must be uninterrupted.
1in type "C" Helical control track level is equiv. to short circuit flux of 100nWb/m.

5. Time code
Must be acc. IEC 461 and recorded on the cue track. (Audio 3 track on 1in type "C".)

The time-code has to be *continuous* and *sync-locked* and run from the *first second of tape lead-in*, through to the *last second of tape lead-out*, preferably in such a way that at the start of the actual programme, the time-code decoder reads

APPENDIX 3

00.00.00.00. or 10.00.00.00. Each address on the tape should be unique and follow on in sequence from that which precedes it.

Note: it is advisable to record the time-code separately, in one pass, so as to ensure perfect continuity.

6. Tape lead-in (Fig 1)
Video
3 min "colourbars". (100/0/100/0 or 100/0/75/0 which must relate to programme content).
Followed by 40 sec "video black".

Audio
During "colourbars", at least 1 min of 1kHz (or 900Hz) pilot tone (and 1 min of Dolby tone if tape is Dolby "A" encoded) at ref.level, on the following tracks:

2 in Quad	If Mono:	Normal Mono Track.
	If Stereo/Dual:	Tracks L and R.
1in type "C"	If Mono:	Track 1.
	If Stereo/Dual:	Tracks 1 and 2.

Ref.level is equivalent to a short-circuit flux of 100 ± 10 nWb/m acc. CCIR REC 469-2.

When using two soundtracks, the two pilot tones must be in phase.

There must be *no sound* during the 40 sec "video black", as this forms part of the programme disc lead-in.

7. Tape lead-out
After actual programme end, at least 30 sec "Video black" and *continuing time-code*, with *no sound* present.

8. Tapes/title
1 tape is required *per programme disc side*. This should comprise:

INTERACTIVE VIDEO

Tape Lead-in .... Actual Programme .... Tape Lead-out.

9. Playing-time
Maximum actual programme time per disc side:

37 min for Active Play (CAV) mode.
58 min for Long Play (CLV) mode.

For material having a duration longer than 1 disc side, it is preferable to divide the playing time equally across disc sides.

10. Actual programme
Start and end (first and last picture) of actual programme have to be stated in time-code on the covering "Master-tape" form. It is imperative that a *continuous control track* is maintained throughout the tape lead-in, actual programme and tape lead-out.

Video
Maximum luminance level is 1.05V. (measured from sync tip).

Colour burst *must* be present throughout, even with B&W productions.

Audio
Maximum audio level (Peak level) is 8db above ref.level and equivalent to a short-circuit flux of 250 nWb/m.
Short-duration speech overshoots of +3db are acceptable, as active limiters are used during disc mastering.

Peak programme meters used during recording should have an integration time acc. IEC 268-10.

Dolby "A" Noise Reduction is recommended.

Audio pre-distortion is not permissible.

Programme sound should be recorded on the following tracks:

APPENDIX 3

2in Quad   Mono: Normal Mono Track.
               Stereo/Dual: Tracks L and R in phase.

Note: on final programe disc:

Stereo Track L will become left channel.
Stereo Track R will become right channel.
Mono Track will be fed to left and right channels.

1in type "C"   Mono: Track 1. (and Track 2 if desired).
               Stereo/Dual: Tracks 1 and 2 in phase.

Note: On final programme disc:

Mono: Track 1 will be fed to left and right channels.

Stereo/Dual: Track 1 will become left channel.
              Track 2 will become right channel.

11.   Vertical interval signals
11.1 LaserVision Cue codes
These codes are generated by the LaserVision Cue-code Inserter and must be present if it is desired to incorporate chapter numbers, or automatic still-frames, on the final disc.

Lines 14, 15 and 19 (and 327, 328 and 332), are reserved for these cue-codes which give information pertaining to:

(a) Field dominance.
(b) Chapter indication.
(c) Automatic still-frame.

Note: When recording LaserVision Cue-codes on to 1in type "C" helical tape format, it is necessary for the VTR to be fitted with the "sync. option" (available from the manufacturer), in order to satisfactorily record the early lines of each field.

INTERACTIVE VIDEO

No other signals are permissible on lines 14, 15, or 19. (327, 328, or 332).

11.2 Teletext
Lines 20 and 21 (333 and 334) are reserved for Teletext signals.

11.3 VITS
Vertical Interval Test Signals are permissible in lines 16, 17 and 18 (329, 330 and 331), however, during mastering, any signals on these lines will be blanked.

12. Master-tape form
Each tape has to be supplied with a covering Master-tape form, on which should be stated:

(a) Programme information — Title and Part (side) Number.
Playing time.
Programme owner.

(b) Video format — Tape standard ie, PAL 625/50.
Tape system ie, 2in Quad, or 1in type "C".
Teletext present. yes/no.
LaserVision Cue-codes present. yes/no.

(c) Audio format — Mono, Stereo, or Dual soundtrack.
Dolby "A". yes/no.

(d) Programme origin — 35 mm, 16 mm, video, or standards-converted.
Colour, or B & W.

(e) SMPTE Time-codes for — Start and end of actual programme.
Chapter cue-codes (if applicable).

APPENDIX 3

		Automatic still-frame cue-codes (if applicable). Start and end of test signals.
(f)	LaserVision Cue-codes	Total quantity of chapter cue-codes present.
		Required preset reading of chapter number at start of disc side.
		Total quantity of automatic still-frame cue-codes present.
(g)	Field dominance	"First", or "Second" field dominance. (Required for Active Play CAV Programmes no containing LaserVision cue-codes).
(h)	Programme transfer mode	Active Play (CAV), or Long Play (CLV).
6I)	Colourbars	Type ie, 100/0/100/0 or 100/0/75/0.

Note: a Master-tape received without an accompanying form cannot be accepted.

147

INTERACTIVE VIDEO

Figure A3.1. Preferred format

TIME-CODE (SMPTE)	VIDEO	AUDIO
23.56.20.00	Colourbars	Pilot tone (and Dolby tone, if applicable) on all tracks with programme sound.
23.59.20.00	Video black	Silence
00.00.00.00	Prog start	Prog start
	PROGRAMME	
xx.xx.xx.xx	Video black (30 secs)	Silence (30 secs)
End	End	End

148